AEROMEDICAL CERTIFICATION EXAMINATIONS

SELF-ASSESSMENT TEST (ACE SAT)

WILLIAM E. WINGFIELD
BS, NREMT-P, FP-C, RN, CFRN, CCRN, RRNA

Printed in the United States of America.

Wingfield, William E.

The aeromedical certification examinations self-assessment test

/ William Wingfield

ISBN 978-0-6151-9124-9

Revision 1.8

Effort has been made to confirm the accuracy of the information presented and to describe generally accepted practices in the field of critical care transport medicine. However, the author and publisher are not responsible for errors, oversights, omissions or for any consequences from application of the information in this book. Neither the author nor the publisher make any warranty, express or implied with respect to contents of this book.

The author and publisher have exerted every effort to ensure that drug selection and dosage set forth in this text are in accordance with current recommendations and practice at the time of publication. However, in view of ongoing research, changes in government regulations and the constant flow of information relating to drug therapy and drug reactions, the reader is urged to check the package insert for each drug they administer in practice.

It is the responsibility of the reader to carefully evaluate the information presented in this text and to apply their practice of medicine in a careful, prudent manner consistent with standards of care as well as individualizing that care to the specific patient at hand.

(ACE SAT)

PREFACE

I created this text in the interest of assisting Paramedics and Nurses in their efforts to prepare for various certification exams as well as to help educate these practitioners in the advanced science, theory and practice of critical care transport medicine. I have focused the material to prepare readers for exams to include the Flight Paramedic Certification (FP-C) offered by the Board for Critical Care Transport Paramedic Certification, the Certified Flight Nurse (CFRN) and Certification for Emergency Nurses (CEN) both offered by the Board of Certification for Emergency Nursing.

For optimal success I would encourage you to prepare for these specific exams by reading the following texts. (This list is short and based on the assumption that the audience reading this book with the intention of attempting certification, has already achieved certification in many of the "industry standard classes" such as BLS, ACLS, PALS, PHTLS, BTLS, TNCC, ATLS, ABLS, AMLS, NRP, CCEMTP etc.);

- *Air & Surface Patient Transport: Principles and Practice* by Renee Semonin Holleran
- *The Manual of Emergency Airway Management* by Ron Walls, M.D.
- *All you Really Need to Know to Interpret Arterial Blood Gases* by Lawrence Martin, M.D.

While this short list of texts does not begin to encompass *all* of transport medicine theory, it is a practical start. Thorough knowledge and understanding of the content in these texts will undoubtedly prepare any clinician for the exams this text addresses. This text specifically was designed to assess your readiness for the exam(s) you plan to take. I am not suggesting that reading these three texts and successfully passing one or more of the certification exams will make you the best patient care provider but I would argue that you will be a *better* patient care provider than you were prior. This text, like any, is a reference and likely contains errors and opinions. The reader should evaluate all ideas presented here in a thorough manner to reach the best-founded decisions in his/her patient care. Ultimately all patient care must be tailored to the specific patient and situation in which that patient is encountered.

At the end of each chapter I have detailed the answers for the questions with the rationale I used to determine the answer selection. Reading of the rationale is important, not just the answer selection, as this offers the physiological reasoning behind the answer as well as test taking tips and strategy. I have spent the vast majority of the last twenty years studying, practicing and teaching transport medicine. I have also spent a large part of the last twenty years reading, writing and most importantly *taking* exams. It is with this text that I hope to relate that experience to you.

Obviously you will disagree with some of my answer selections or may feel that some questions have more than one correct answer. Hopefully, those questions will have your concerns alleviated in the rationale section. Medicine, more specifically transport medicine is an ever-growing and changing field of science. It is difficult to present a text that is completely current upon publication. However, one must remember that the certification exams are limited to almost the same 'lag time' in terms of updating questions. Standards of care should be, and typically are the tested material, as such, tomorrow's landmark study on topic 'X' will not likely be tested as a standard of care for quite some time.

This exam review book does ***not*** *present the actual questions from the target examinations*. While access to actual questions may be available somewhere, it is not in the best interest of you, this author or the thousands that have worked so hard to obtain these certifications to provide those specific questions to you. Should you feel so inclined as to share actual test questions you should be warned. The exam administrators prohibit such activity as well as their certifying bodies and ultimately it is unethical. I encourage anyone to contact me with text suggestions but please **do not** send me actual test questions. I have been involved in one way or another with the FP-C exam since it's earliest days and have attempted to stay current on the make up of the exam's writing team, motivations, politics and tactics. Having taught an FP-C/CFRN review course consistently for the last five years, I have seen numerous candidates succeed in achieving certification. In my opinion, the exam questions in this text are very similar to the questions you will encounter on board exams. I feel confident that should you be comfortable with the material presented here, you should have no problem passing any of the exams targeted. More importantly, if you are proficient with the material presented here you should possess the core knowledge to be a quality practitioner. That is what these certifications are intended to "certify".

Paige…

You are my everything.

Dad…

Thank you for teaching me to take pride in my work.

Those that took time to teach me…

To all of you, my sincerest thanks. I can never repay all that I owe.

TABLE OF CONTENTS

(Total of 560 questions)

FLIGHT PHYSIOLOGY and AVIATION STANDARDS

1. Barotitis media typically occurs upon;
   a. ascent
   b. descent
   c. both
   d. neither

2. Barobariatrauma is best prevented by;
   a. positive pressure ventilation
   b. pre-oxygenation
   c. PEEP or CPAP usage
   d. none of the above

3. Your portable O_2 tank showed 1100psi on the pressure regulator during the shift change check at 0630hrs. The tank was left exposed to the sun and two hours later shows a pressure of 1350psi. This is an example of;
   a. Boyle's Law
   b. Charles's Law
   c. Gay-Lussac's Law
   d. Graham's Law

4. Your patient is reportedly experiencing seasonal allergies in addition to the gigantic heart attack he's being transported for. Barosinusitis may occur during transport. You would expect the patient to complain of barosinusitis symptoms during;
   a. ascent
   b. descent
   c. both
   d. neither

5. You are currently loading a patient at an airfield, which is at sea level. The barometric pressure is currently 760torr. Your patient is on supplemental O_2 with an FiO_2 of 0.3. The partial pressure of the oxygen as it enters the patient's nasopharynx is;
    a. 228mm Hg
    b. 21%
    c. 166mm Hg
    d. Impossible to determine from the information given

6. During ascent you note increased sporadic bubbling in a patient's chest tube drainage system. This may demonstrate;
    a. Boyle's Law
    b. Henry's Law
    c. Gay-Lussac's Law
    d. Graham's Law

7. Boyle's Law will affect all but which of the following;
    a. a plaster cast
    b. a foley catheter balloon
    c. an endotracheal tube cuff
    d. an intra-aortic balloon pump balloon

8. You lifted from an airfield at 2100'MSL, 98°F with winds at 20kts with a heading of 110°. Your helicopter has climbed to 850'AGL with a rate of ascent of 300'/min. Your current altitude is;
    a. assigned by the tower operator
    b. 850' MSL
    c. 2950' MSL
    d. 2050' MSL

9. The emergency locator transmitter frequency is;
    a. 7700Hz
    b. 7500Hz
    c. 240Hz
    d. 121.5Hz

10. You should anticipate a greater change in atmospheric pressure per altitude change when flying;
    a. near the equator
    b. over large bodies of water
    c. near the north pole
    d. during night time hours

11. Charles's Law states that as you increase the temperature of a gas you should anticipate;
    a. a pressure increase
    b. a volume increase
    c. a solubility increase
    d. a reduced gas weight

12. Gay-Lussac's Law most closely resembles;
    a. Boyle's Law
    b. Dalton's Law
    c. Graham's Law
    d. Charles's Law

13. The release of gas exhibited when opening a carbonated drink is a clear demonstration of which gas law?
    a. Boyle's Law
    b. Dalton's Law
    c. Henry's Law
    d. Graham's Law

14. Barodontalgia will likely be exacerbated by;
    a. ascent
    b. descent
    c. turbulent flight
    d. increased cabin pressurization

15. As you ascend you would expect the pressure in your patient's pneumatic anti-shock garment (PASG, aka-MAST) to;
    a. increase
    b. decrease
    c. stay the same
    d. fluctuate slightly with a descending trend in pressure

16. "At a constant pressure, the volume of gas is directly proportional to the absolute temperature of the gas" describes;
    a. Boyle's Law
    b. Dalton's Law
    c. Gay-Lussac's Law
    d. Charles's Law

17. For every 100 meters climb in altitude you can anticipate;
    a. decreasing concentration of oxygen of 1%
    b. a temperature drop of 1°C
    c. increasing concentration of nitrogen of 0.3%
    d. a temperature increase as you get closer to the sun

18. Graham's law is influenced more by;
    a. velocity of gas flow
    b. gas density
    c. electrical gradient of the gases in question
    d. physiologic properties of the gas in question

19. Dalton's Law demonstrates that the concentration of oxygen at 40,000ft MSL should be;
    a. 12%
    b. 16%
    c. 18%
    d. 21%

20. Which of the following is true concerning air at high altitudes?
    a. colder than air at lower altitude
    b. more moist than air at lower altitude
    c. higher concentration of nitrogen than air at lower altitude
    d. lower concentration of carbon dioxide than air at lower altitude

21. Which of the following is a stressor of flight
    a. barometric pressure changes
    b. vibration
    c. noise
    d. all of the above

22. Which of the following is not a factor influencing the stressors of flight?
    a. alcohol
    b. tobacco use
    c. thermal changes
    d. hypoglycemia

23. One 'standard atmosphere' weighs;
    a. 760torr
    b. 14.7psi
    c. 101.3kPa
    d. all of the above

24. An altitude of 45,000ft MSL would be considered;
    a. physiologic zone
    b. space-equivalent zone
    c. physiologic deficient zone
    d. space

25. An otherwise healthy individual exposed to the ambient environment at 20,000ft MSL would experience;
    a. hypoxic hypoxia first
    b. stagnant hypoxia first
    c. hypemic hypoxia first
    d. histotoxic hypoxia first

26. A patient with an H&H of 6 & 18 with an SaO_2 of 95% will suffer primarily from __________ if taken to 3500ft MSL from sea level with O_2 via a non-rebreather mask at 10lpm.
    a. hypoxic hypoxia
    b. stagnant hypoxia
    c. hypemic hypoxia
    d. histotoxic hypoxia

27. A pilot reports that he's experiencing dizziness while ascending to a high altitude in an unpressurized fixed-wing aircraft. He would be considered in which stage of hypoxia?
    a. indifferent stage
    b. compensatory stage
    c. disturbance stage
    d. critical stage

28. You are transporting an infant with a congenital heart defect from New Orleans to Santa Fe. The altitude and weather indicate your barometric pressure will change from 760mm Hg to 605mm Hg. Your patient's orders read that you must strictly maintain the FiO_2 at exactly 30%. Without an oxygen sensor/analyzer in use, you calculate the delivered FiO_2 upon landing in Santa Fe should be;
    a. 30%
    b. 32%
    c. 36%
    d. 38%

29. You are flying in a Learjet at 40,000ft MSL when you experience an 'explosive decompression'. You should anticipate your time of useful consciousness is;
    a. 30-60sec
    b. 60-90sec
    c. 120-150sec
    d. 3-5sec

30. You will be transporting a hemodynamically stable patient with pneumocephalus. Which of the following would be the best transport choice.
    a. Ground CCT unit, transport time 2hrs 20min, change in altitude of 1000ft as you cross the mountain pass from sending to receiving facility.
    b. Rotor wing, transport time 1 hour, change in altitude of 6000ft to safely cross mountain range that night.
    c. Fixed wing (pressurized), transport time 1hr 15min to closes airfield, 40min by air with 1hr from airfield to destination facility.
    d. Horseback drawn sled, transport time 8 days, change in altitude dependent upon how many times the patient falls off the sled.

31. Commonly, healthy, unimpaired flight crewmembers will begin to experience deterioration in night vision at;
    a. 500ft AGL
    b. 500ft MSL
    c. 5000ft AGL
    d. 5000ft MSL

32. Night vision is primarily a function of the __________ which are located ___________ in the retina.
    a. rods , center
    b. rods, periphery
    c. cones, center
    d. cones, periphery

33. Scuba Steve has descended 66ft below the surface in a lake. At that point he is experiencing;
    a. two atmospheres of pressure
    b. three atmospheres of pressure
    c. one atmosphere of pressure
    d. no atmospheric pressure as he's submerged below the air surface

34. Techniques commonly utilized to relieve the discomfort of barotitis media include all but which of the following;
    a. valsalva maneuver
    b. positive pressure mask ventilation
    c. slower altitude change
    d. needle decompression of the tympanic membrane

35. When administering high concentration oxygen to alleviate hypoxic hypoxia you are altering which component of which gas law?
    a. solubility; Graham's Law
    b. solubility; Henry's Law
    c. partial pressure; Henry's Law
    d. partial pressure; Boyle's Law

36. You have just experienced a hard/crash landing. You should rendezvous with your crew;
    a. at the 12 o'clock position
    b. at the 3 o'clock position
    c. at the 9 o'clock position
    d. at the 6 o'clock position

37. The proper shut-down procedure for a typical aeromedical aircraft commonly follows the priority of;
    a. battery, fuel, oxygen, rotor brake
    b. fuel, throttle, battery, rotor brake, oxygen
    c. throttle, battery, fuel, rotor brake, oxygen
    d. throttle, fuel, battery, rotor brake, oxygen

38. Critical phases of flight do not include;
    a. take off & landing
    b. changing heading
    c. changing altitude
    d. all of these are critical phases of flight

39. Your priority post crash landing is;
    a. acquiring water
    b. going for help
    c. finding shelter from the environment
    d. assessing aircraft damage

40. Cellular phone use in flight is;
    a. a violation of FAA guidelines
    b. a violation of FCC guidelines
    c. will potentially interfere with aircraft navigation systems
    d. will overwhelm the cellular tower sites

41. Flight crewmembers should notify the pilot of any potential hazards by;
    a. shouting "Look out!" while pointing frantically
    b. announce the hazard, location and heading
    c. announce the location, hazard and heading
    d. announce the heading, location and hazard

42. The FAA requires that crewmembers observe sterile cockpit during;
    a. take off
    b. landing
    c. taxiing
    d. all critical phases of flight

43. Per FAA guidelines an aeromedical program may only fly;
    a. VFR in VMC
    b. IFR in VMC
    c. VMC in IMC
    d. IFR in VFR

44. Regarding your aircraft's ELT;
    a. it should activate upon an impact of four G's
    b. it may be manually activated
    c. it may not be activated upon a crash landing
    d. all of the above are accurate

45. Standard night LZ recommendations include all but which of the following;
    a. communication with ground personnel
    b. 100ft X 100ft LZ area
    c. all available personnel and rescue apparatus should focus all lighting to the center of the LZ
    d. illumination of power line pole tops in the area

46. Your pilot states that visibility is being reported as "500 and 2". This indicates;
    a. there is a 500 mile visibility with a 2 mile ceiling
    b. there is a 500 foot ceiling with a 2 mile visibility
    c. there is a 500 mile visibility with two storms 'visible' in the area
    d. there is a 500 foot ceiling and a 2000 foot ceiling above that

47. You pilot informs you that you will be landing on runway two-six. Your heading upon final approach should be;
    a. 80°
    b. 26°
    c. 180°
    d. 260°

48. The best survival strategy post crash would be;
    a. teams of two should begin attempting to go for help and/or locate assistance
    b. uninjured personnel should immediately light the wreckage on fire to signal would be rescuers
    c. remove the dead from the primary area creating a temporary 'morgue' area
    d. eat the pilot

49. Your pilot has asked you to come to the flight deck and sit in the empty co-pilot seat while you descend to your destination airfield. He state's he needs you to notify him when you see the ground through the current solid cloud cover. He is;
    a. suffering from loss of night vision and this situation is critical
    b. attempting to prevent spatial disorientation by focusing on his gauges
    c. making a poor effort at humor and actually scaring his crew inappropriately
    d. attempting to avoid a critical mishap without alerting the crew fully to the situation

50. Your rotor-wing pilot informs you he must make an emergency landing due to a complete power failure and to prepare. You should anticipate;
    a. an immediate descent while he performs an auto-rotation maneuver
    b. audible and visual alarms to activate suddenly and repeatedly
    c. calling "MAYDAY! MAYDAY! MAYDAY!" with your call-sign and location immediately
    d. all of the above

KEY & RATIONALE

FLIGHT PHYSIOLOGY and AVIATION STANDARDS

1. B Barotitis media related pain is the discomfort caused by pressure in the middle ear not being in equilibrium with the atmosphere. During descent the atmospheric pressure pushing in on the tympanic membrane increases. If that same pressure cannot access the middle ear via the eustachian tubes an imbalance occurs and pain will result. It is uncommon for air to be unable to escape the middle ear via the eustachian tubes so problems on ascent are less common.

TEST TIP
barotitis media ⇒ descent problem

2. B Barobariatrauma is caused by a large/sudden release of nitrogen from the adipose tissues of the body upon decompression. Normally the body would deal with this release of nitrogen by simply 'blowing off' the nitrogen via the pulmonary system just like carbon dioxide. However, in the very obese patient, the combination of excess nitrogen stores in addition to decreased functional residual capacity (FRC) and tidal volumes (Vt) lend to an inability to remove the nitrogen fast enough. By thoroughly pre-oxygenating, you replace the nitrogen in the FRC with oxygen therefore creating a larger nitrogen partial pressure gradient. This increased gradient allows for more rapid transfer of nitrogen from the plasma to the alveoli.

TEST TIP
barobariatrauma ⇒ pre oxygenate first

3. C Gay-Lussac's Law is very similar to Charles's Law in that they both relate temperature to changes in pressure or volume respectively. In this case the volume is held constant (by the metal cylinder) and thus pressure is changed.

TEST TIP
"Gay-Lussac's is Charles's gay brother"

4. A Barosinusitis refers to air trapped in the sinuses. During ascent the falling atmospheric pressure can't keep the air in the sinuses compressed, thus that gas attempts to expand causing pressure on the sinus wall and pain. As the patient descends, the increased atmospheric pressure will assist with controlling the barosinusitis.

5. A Partial pressure is calculated by multiplying the given gas concentration by the total pressure present. Thus 760mm Hg X 0.3 = 228mm Hg.

6. A The increased sporadic bubbling suggests air is escaping the patients chest. Based on the assumption the chest tube was placed for a pneumothorax, it would stand to reason that a residual pneumothorax is attempting to expand with the climb in altitude (Boyle's Law) and thus some air is escaping via the chest tube. This sporadic bubbling could also infer an increased air leak in the chest or a chest drain system leak but these were not offered as answer choices.

7. B Boyle's law refers to gases attempting to expand as outside pressure is reduced. Gases in tissues will cause expansion of tissues under a cast to expand with ascent. This principle obviously applies to endotracheal tube cuffs as well as the balloon utilized with an intra-aortic balloon pump. The foley balloon is filled with water, which will not expand for all intensive purposes, regardless of pressure changes encountered during transport. Any cast less than seven days old should be bi-valved and secured with an elastic bandage prior to transport. Monitoring of the distal extremity during transport is indicated with loosening of the elastic bandage should circulation be compromised. Reporting of this cast bi-valving should occur at the receiving facility for appropriate replacement.

TEST TIP
"Boyle's balloon"

8. C This question has an enormous amount of useless information or distracters. The terms MSL and AGL must be understood to determine the appropriate answer. MSL or "mean sea level" refers to the elevation above sea level. AGL or "above ground level" refers to the elevation above the ground directly below the aircraft. Thus beginning at an altitude of 2100ft MSL and climbing 850ft AGL would make your altitude 2100 + 850 or 2950ft MSL.

9. D The emergency locator transmitter or "ELT" is a device attached to an aircraft which will automatically send out a radio signal when triggered. The device is located and designed to be automatically triggered by an inertial change of 4 G's such as in a "hard landing" or crash. This device can be triggered manually from the cockpit as well and via a toggle switch on the device itself. ELT's are commonly DOT orange or bright yellow in color, approximately 10 inches long and 4 inches square. (see picture left) They are required to have a self-contained battery and antenna to enable physical separation from the aircraft while continuing to operate properly. NOTE: Civilian ELT's historically used the 121.5Hz frequency while the U.S. military uses a frequency of 243Hz (243=121.5 X 2). Newer ELT's and satellite monitoring will be utilizing a 406Hz frequency range which may be mandated around 2009 when specific satellites for the current system cease to function.

10. C One must remember that the atmosphere is held to the Earth by gravity and all laws of physics apply including centripetal force. The Earth's rotation causes the atmosphere to be 'flung out' around the equator (aka- equatorial bulge) where the rotational speed is maximal versus the poles. As such the atmosphere is physically thicker at the equator (over land masses, water is subject to the same equatorial bulge). When changing altitude at the poles you will be rising through a greater *overall percentage* of the atmosphere at that point, thus larger changes in atmospheric pressure will be experienced. Large bodies of water may affect humidity and thus density altitude. Night flying may also alter density altitude due to temperature changes.

11. B Charles's Law states that changes in temperature will cause a change in volume assuming pressure is constant. Thus as you heat the air in a balloon you would anticipate the gas to expand (assuming the balloon provides no significant resistance to such).

TEST TIP
"Charles's Centigrade"

12. D Gay-Lussac's Law closely mirrors Charles's Law. Gay-Lussac's Law states that as you increase the temperature of a gas in a fixed volume container, the pressure must rise. Charles's Law: Temperature increase results in a volume increase if constant pressure is maintained. Gay-Lussac's Law: Temperature increase results in a pressure increase if constant volume is maintained.

13. C Henry's Law states; At a constant temperature, the amount of a given gas dissolved in a given type and volume of liquid is directly proportional to the partial pressure of that gas in equilibrium with that liquid. Once you open the bottle top, the pressure above the fluid immediately equalizes with the atmosphere becoming lower than the pressure of the gas dissolved in the beverage. As such, you begin to witness the gas escape the liquid in an attempt to reach equilibrium with the outside environment. The rate of this escape is governed by the partial pressure difference, the solubility of the gas in the liquid as well as the surface area which the atmosphere and liquid interface. Hence the narrow bottle neck style used in many carbonated beverages.

TEST TIP

"Henry's Heineken"

14. A Barodontalgia refers to gas trapped between dental appliances and teeth. This gas will attempt to expand upon ascent but due to the dental appliance trapping, can only place pressure on the tooth underneath and subsequently any exposed nervous tissue. It has been reported that in extreme cases this can be significant enough to disrupt the dental appliance fixation.

15. A As you ascend, the atmospheric pressure will drop allowing the gas in the PASG to attempt to expand. Because the PASG is designed to maintain a maximum external girth (fixed container size), the expanding gas will translate to increased pressure inward and will be indicated by an elevated pressure gauge reading or "pulling" of the Velcro attachments.

16. D This is Charles's Law.

17. B As you climb in altitude the temperature drops. Anticipate a 1°C for every 100 meter climb. *Concentration* of oxygen does not change with altitude change. There is 21% oxygen at sea level, on top of Mount Everest and at 50,000 ft MSL. There are fewer molecules of *all* gases as you go up but 21% of those molecules are still oxygen. D would be a good choice if we were asking about the Greek mythological character Icarus.

TEST TIP
Climb 100m = Drop 1°C

18. B Graham's Law states; The rate of diffusion of a gas is inversely proportional to the square root of its molecular weight. Translation: dense gases diffuse slower than less dense gases through liquid. The velocity of the gas, it's electrical charge and physiologic properties all may effect your patient in some manner but none pertain to Graham's Law.

19. D Dalton's Law states; The total pressure exerted by a gaseous mixture is equal to the sum of the partial pressures ("the entire gang") of each individual component in a gas mixture. As we go up in altitude we find that the percentage concentration does not change, the atmosphere is uniform in its molecular distribution. A sample of gas from 40,000ft MSL will have 21% oxygen, 78% nitrogen and 1% trace gases; identical to a sample at sea level. However, because the atmospheric pressure at 40,000ft is substantially less we find that when calculating the partial pressure of that oxygen (PO_2) it's substantially less at altitude vs. sea level. Using the calculation; PO_2 = Atmospheric pressure X FiO_2 we can apply this to the altitudes in question as;

PO_2 = 760torr X .21 = 159.6torr (sea level)
PO_2 = 162torr X .21 = 34.1torr (~40,000ft MSL)

TEST TIP
"Dalton's Gang"

20. A As you ascend the air's temperature will decrease and it will become dryer. Concentrations of the various gases do not change, only their partial pressures.

21. D All of these are considered stressors of flight along with decreased partial pressure of oxygen, thermal changes, decreased humidity, fatigue and G-forces. Know these stressors of flight!

22. C Thermal changes are considered an actual stressor of flight. All the others exacerbate the stressors of flight. One way to approach this question is look at which answer is not like the others. C is the only option that cannot be changed by the flight crewmember.

23. D One "standard atmosphere" is the weight of the atmosphere at sea level at 59°F.

24. C Physiologic zone spans from sea level to 10,000ft MSL. Physiologic deficient zone spans 10,001ft MSL to 50,000ft MSL. Space-equivalent zone spans 50,001ft MSL to 250,000ft MSL. Space is considered everything beyond 250,000ft MSL. (In the United States, anyone ascending more than 50nm or 303,805ft MSL is officially an "astronaut". That's not on the test.. so far.)

25. A Hypoxic hypoxia would result from exposure to an excessively low partial pressure of oxygen. Stagnant hypoxia requires the failure to move oxyhemoglobin to the needed areas of the body thus suggesting a cardiovascular impairment. Hypemic hypoxia would require a deficiency in oxygen carrying capability such as anemia. Histotoxic hypoxia would require some form of oxygen loading or unloading problem like cyanide poisoning.

26. C Hypemic hypoxia refers to an inability to carry oxygen that is available at the alveolar capillary interface. It is commonly practiced that an H&H of less than 8 & 24 should probably receive blood prior to being taken to altitude, with or without supplemental oxygen. This is somewhat consistent with the American College of Surgeons (ACS) recommendation that blood transfusion for a Hgb between 6 and 10 is based on clinical circumstances to include current hemorrhage control, surgical status, care in progress, etc. The ACS suggests that a Hgb<6 almost always should have PRBC's transfused and a Hgb>10 seldom requires transfusion. One can see the influence of hemoglobin on oxygen delivery when the oxygen content (CaO_2) equation is utilized.

$$\mathbf{CaO_2 = (1.34 \times Hgb \times SaO_2) + (PaO_2 \times .003)}$$

Looking at this equation demonstrates that a drop in Hgb from the normal of approximately 15g/dL to 7.5g/dL decreases the oxygen carrying capacitance (CaO_2) almost 50% without regard to actual PaO_2. This drop in Hgb cannot be realistically compensated with an increase in FiO_2 alone.

27. C The disturbance stage follows the indifferent and compensatory stages. It is typified by dizziness, sleepiness, tunnel vision and cyanosis. Without rapid recognition and correction of this, progression to significant disturbance and finally critical stage are eminent with an aircraft incident/accident very likely.

28. D The formula utilized for this calculation is;

$$(P_{B1} \times FiO2_1)/P_{B2} = FiO2_2$$

P_{B1} is the barometric pressure at the starting point of the transport. $FiO2_1$ is the FiO_2 being delivered at the starting point of the transport. P_{B2} is the barometric pressure for the destination and $FiO2_2$ is the FiO_2 that must be delivered at the destination's barometric pressure to ensure the same partial pressure of oxygen.

TEST TIP

$$(P_{B1} \times FiO2_1)/P_{B2} = FiO2_2$$

29. D The term 'explosive decompression' refers to any decompression of a pressurized vessel which takes less than 0.1 second to occur. The term comes from the rapid equilibration causing a sound wave that sounds like an explosive bang. This coupled with immediate decompression and cooling of the aircraft cabin causing water vapor in the air to immediately condense forming fog lending to the appearance of smoke. Rapid decompression is a similar event but takes longer. The time of useful consciousness (TUC) tables demonstrate decreasing amounts of time with the greater the decompression and when explosive, these times are reduced by 50% versus rapid decompression TUC. You should not devote any real effort to memorizing these tables, instead understand the principles behind them and you can realistically assume that any scenario like the one above will be keyed for the shortest time option given.

ALTITUDE	TUC*
18,000ft	20-30min
22,000ft	10min
25,000ft	3-5min
28,000ft	2.5-3min
30,000ft	1-2min
35,000ft	30-60sec
40,000ft	15-20sec
43,000ft	9-12sec

**-times based on controlled decompression of subject at rest*
Explosive decompression can reduce times by 50%
Exertion at time of decompression reduces time dramatically also

30. A The patient's condition (pneumocephalus) should alert you to the risks associated with changes in altitude. Climbing this patient to altitude will invoke Boyle's Law to expand the pneumocephalus causing increased ICP's. While the transport time would be shorter, rotor-wing would be more hazardous. Fixed wing with careful attention to cabin pressure may be adequately safe but total time is significantly longer in this case.

31. D The eyes are especially susceptible to hypoxia. The otherwise healthy individual begins to experience vision changes due to hypoxia at approximately 5000ft MSL. You must recognize that if you work from a base which sitting on the ramp is at 4000ft MSL, you only need to climb 1000ft AGL to begin losing night vision. If you are walking to your aircraft in Santa Fe, New Mexico at approximately 7000ft MSL, you would need to dig a hole 2000ft deep to begin regaining your night vision, and you haven't even boarded the aircraft yet!

32. B The rods, located along the periphery of the retina are primarily responsible for night vision. It is rhodopsin, a pigment in the rods that changes shape when absorbing light that provides us with night vision. This pigment bleaches almost instantaneously when struck by white light and requires up to 30 minutes to completely return to the pre-exposure levels. It is because of the rods peripheral placement that 'scanning' is the recommended method of observing for hazards during night operations. By scanning you are more likely to see objects in your peripheral vision than if you look directly at them. Hence the guidance to look 'next to' any particular object you which to observe at night focusing your mental image on your peripheral vision. Cones are designed for daytime lighting and because we are predominantly a diurnal creature they have developed in the central area of our retinas.

33. B The diver will experience an additional atmosphere of pressure for every 33 feet dived. This alters somewhat with fresh vs. salt water but for testing purposes will not be significant. What will prove significant is exactly how the question reads. Had the question asked, "How many atmospheres of *water* pressure were on the diver?" Your answer would then be, "Two atmospheres."

TEST TIP
33 feet water = 1 atmosphere

34. D Be careful, this question is asking which technique is <u>not</u> *commonly utilized*. All of the techniques above D are common and relatively benign management strategies. The last can be employed but it is not *common*.

35. C Increasing the FiO_2 manipulates the partial pressure gradient between the alveolar oxygen and plasma arterial oxygen tensions. The gas law most influenced in this practice would be Henry's Law.

36. A Standard crew survival, post-crash procedures dictate you should rendezvous with your team at the 12 o'clock position unless that position is deemed unsafe. Your subsequent choices would then be 3, 9 and finally 6 o'clock positions.

37. D The throttle mechanism will reduce fuel availability to the combustion part of the power plant but these commonly require electrical power to allow their mechanisms to respond to user inputs. The fuel shut-offs typically have a similar design and again require electrical power to respond. Rotor brake application and oxygen shut-off are helpful but are not always applicable/practical, their order is really much less important and likely more situation dependent, (Do you still have an active rotor system above you? Do you have an active fire in progress? etc.)

Your priority is: 1st throttle 2nd fuel 3rd battery

Some aeromedical platforms are equipped with automatic fire suppression systems, do not worry about these for purposes of the exam as they are not available on the majority of aircraft utilized for aeromedical transport in the United States at this time.

TEST TIP

Hard/Crash Landing ⇒ Throttle, Fuel, Battery

38. D The easiest way to remember this point is that a critical phase of flight basically includes everything *except straight level flight*.

39. C Shelter from the elements is your immediate priority. This includes procuring fire to maintain warmth in a cold environment. ***Never*** leave the crash site, Search and Rescue (SAR) will find the wreckage before they find you wandering around. Water becomes a higher priority after 24 hours without rescue. Assessing the aircraft for salvageable/usable survival supplies is indicated once deemed safe to do so.

40. B The FCC guidelines prohibit the use of cellular phones in flight. The urban legends of cellular phones causing aircraft to navigate off course or interfere with them are completely false, especially with today's technology and aircraft equipment. Your cellular phone, should it be used while airborne will potentially be received by multiple cellular tower sites but this does not overwhelm their system. It does however make it difficult for them to triangulate your location to identify if you're in your "local area" or "out of network" and this complicates their ability to legally bill you. If you disagree, try leaving your phone 'on' in your pocket next time you fly commercial and see if you can get your flight to land in the wrong city. I keep trying, it never works.

41. C You should announce a hazard by it's location first using the clock system. This alerts the pilot and crew to begin looking that direction as you complete the hazard description. For example, "10 o' clock, aircraft 5 miles, same flight level, closing"

42. D FAA regulations require flight personnel "observe sterile cockpit procedures during all critical phases of flight". As such the only conversation going on should be relevant to the safe operation of the aircraft at that time. Answers A, B & C are all correct but incomplete. The best answer is D. This type of question is commonly placed near the end of the exam where you are mentally fatigued. The less than alert test taker will quickly grab the first answer that triggered a correct response in their mind and never makes it to answer D. This also happens in the test taker that is overly concerned about time limitations and hurrying.

43. A The only option that is true for all aeromedical programs at all times is option A, VFR in VMC. VFR or visual flight rules can be applied to aircraft that are flying in visual meteorological conditions or VMC. FAA guidelines will not permit you to legally fly VFR in instrument meteorological conditions or IMC. Obviously you must fly by instrument flight rules or IFR in IMC. Can you fly as per option B, IFR in VMC? Yes. However the question asks "...an aeromedical program may *only* fly;" Many rotor-wing aeromedical programs are "VFR programs" dictating that they are not allowed to knowingly initiate a flight, which would require IFR for safe completion of the mission. So for purposes of the exams; remember, you <u>may</u> fly VFR *only* in VMC. You *must* fly IFR in IMC (if you're an approved IFR program). You may *never* fly VFR in IMC regardless of your program's approvals, equipment or personnel credentials.

44. D Emergency locator transmitters or ELT's are designed to automatically begin an alert signal upon a crash landing. By design they should automatically trigger with an impact of 4G's however if the impact is at an oblique enough angle to the inertia switch of the ELT, it may fail to activate. ELT's are also designed to be manually triggered, typically by a switch in the cockpit or by a toggle switch on the ELT itself. The device is designed to function from an internal battery as well as an internal antenna. The device can be removed from the aircraft completely and should continue to operate correctly.

45. C Questions requesting you identify *"all but which"* can be confusing and in older testing models frequently were placed at the end of exams where the candidate was anticipated to be mentally fatigued. Option C poses a safety risk to the safe landing of the aircraft. Ground crews should be instructed to divert high-power lighting away from the pilot for safest landing of the aircraft. Options B and D are true recommendations but not considered a '*requirement*' for a night LZ. You should consider option A, a '*requirement*' for safe night landing.

TEST TIP

"Direct communication with ground crews is a <u>requirement</u> for a night LZ"

46. B When reporting visibility this is typically stated with the ceiling followed by the visibility. Flight minimums of the program will dictate if the current conditions allow for safe flight per policy.

Crash with fatalities. Roswell, New Mexico -2001

ELT visible top center picture at 9 o'clock from Oxygen tank. New ELT. Did not activate on impact. Tested perfect post crash.

47. D Runway designations are based upon the heading of the aircraft landing on said runway, less the '0' at the end of the heading. So runway '26' would require a heading of 260° to be perfectly lined up. Landing on the same runway from the opposite direction would require a heading of 080° and hence the runway label at the other end would be '08'. Landing on runway '26 Right' would tell you that there are two runways parallel to one another, both with the alignment of heading 260°. You are being told to land on the runway that will be on your right from that approach heading. If you review the diagram of the airfield below you can see that there are two pairs of runways. The pair on the upper left, if approached from the bottom of the map would require a heading of 360° as indicated by the runway designations "36". The pilot would be instructed to land on either runway 36R (right) or 36L (left). You can see if the approach was from the top of the map the heading would need to be 180° to be lined up and again you would receive instructions for 18R or 18L.

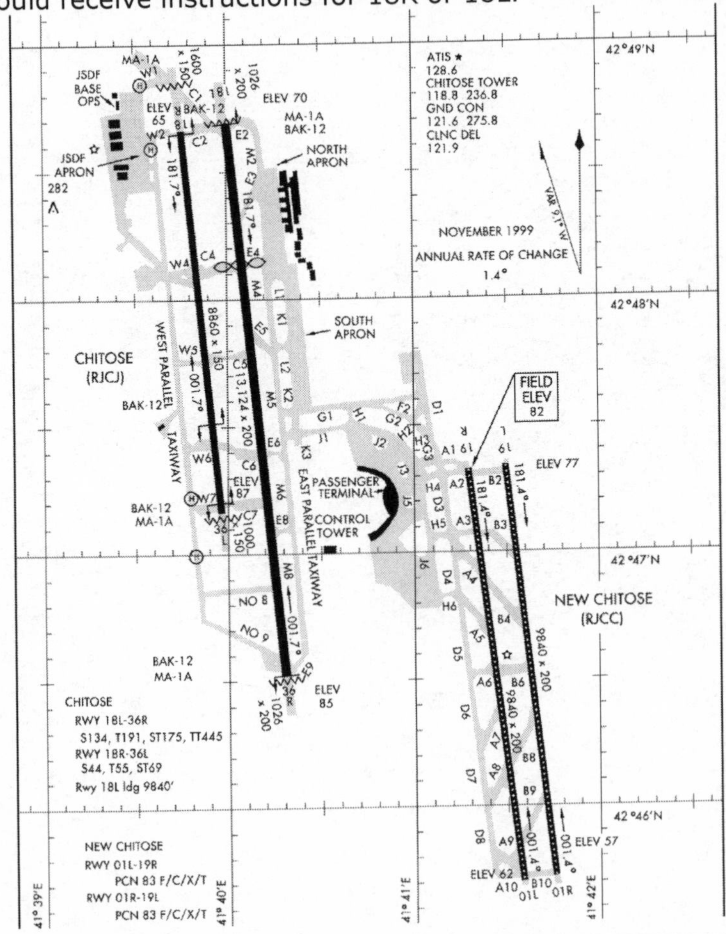

48. C Leaving the wreckage is a cardinal no-no. Search and Rescue (SAR) teams locate wreckage long before they locate wandering crewmembers. Stay put! Generally speaking you don't want to light the wreckage on fire, this simply places you at increased risk for injury, wastes any additional survival resources present in the aircraft and is just overall a bad idea. Removing the dead from your immediate area is appropriate for both physical, mental and emotional health reasons. Long –term survival requires a strong mental state, something that is very difficult to maintain sitting around looking a deceased crewmates. Don't eat the pilot, they're generally sinewy and lack taste.

49. B This is a common strategy for avoiding spatial disorientation which can occur as the aircraft emerges from IMC to VMC conditions. The theory behind this is that should you become spatially disoriented, your situation will not affect the safe operation of the aircraft.

50. D A complete power failure will necessitate an immediate emergency landing. The autorotation maneuver is performed to maintain rotor speed and provide the needed power at the point of landing. This maneuver requires a continuous, controlled descent allowing the air to effectively 'windmill' the main rotor system. While this maneuver is employed you will undoubtedly hear and see multiple alarms, rotor speed horns and the like. Stay calm, verify you've performed your pre-crash duties and ask the pilot for additional instructions. Calling "MAYDAY!" may seem extreme initially but it's always easier to cancel a premature mayday when the emergency is over than to not have called it at all, and be forced to wait for the time lapse required by a PAIP before a search is initiated.

President Bill Clinton's motorcade EMS escort
El Paso International Airport 1997

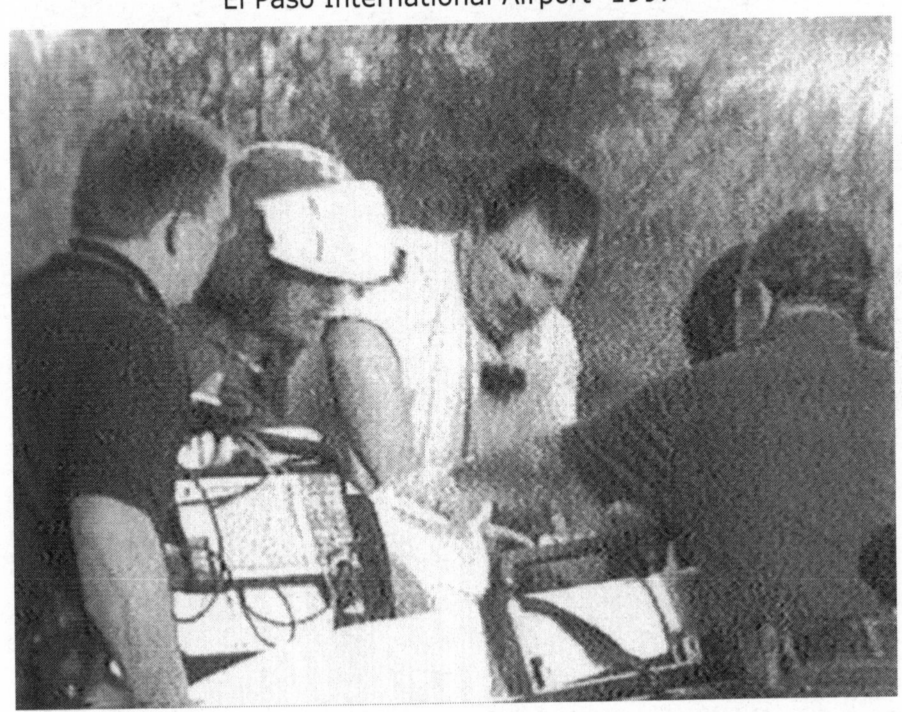

AeroCare scene call
Lubbock, Texas - 2003

RESPIRATORY DISEASE, ACID–BASE DISTURBANCES & VENTILATOR MANAGEMENT

1. The fastest physiological mechanism for acid base correction is;
   a. phosphate buffer system
   b. respiratory system
   c. renal system
   d. bicarbonate buffer system

2. pH is a calculation of the presence of;
   a. potassium ions
   b. carbon dioxide
   c. phosphate ions
   d. hydrogen

3. During aerobic metabolism cells primarily produce;
   a. adenosine
   b. adenosine monophosphate
   c. adenosine diphosphate
   d. adenosine triphosphate

4. The primary metabolic byproducts of completed aerobic metabolism are;
   a. CO_2 and lactate
   b. lactate and water
   c. lactate and pyruvate
   d. CO_2 and water

5. The Bohr effect;
   a. occurs as a result of CO_2 binding with Hgb causing decreased affinity of hemoglobin for oxygen
   b. causes a left shift in the oxyhemoglobin dissociation curve
   c. occurs primarily in the pulmonary circulation
   d. is increased in the presence of carboxyhemoglobin

6. Your 32 year old patient is exhibiting signs of hypoxia to include altered level of consciousness. His SaO_2 is 98% on 4l/min via nasal cannula breathing at a rate of 28/min. His H&H is 6 & 17. His blood pressure is 102/66 with a heart rate of 128/min, his skin is pale but dry and warm overall. He has produced urine at 0.5ml/kg/hr for the last two hours. He is most likely suffering from;
   a. hypoxic hypoxia
   b. histotoxic hypoxia
   c. hypemic hypoxia
   d. stagnant hypoxia

7. The most definitive test for 'shock' would be;
   a. positive D-dimer
   b. MAP<60mm Hg
   c. elevated lactate
   d. decreased potassium

8. Your multi-system trauma patient has been receiving large amounts of crystalloids and packed red blood cells (PRBC's) as part of an ongoing resuscitation. He remains 'shocky', core temp has been maintained within normal limits. The patient is speculated to have ongoing abdominal bleeding complicated by a pneumothorax (which has been treated by tube thoracostomy) and a large pulmonary contusion. Based on the limited information provided, you would anticipate his oxyhemoglobin dissociation curve has shifted;
   a. right
   b. left
   c. up
   d. down

9. While in the ICU for a GI bleed, the patient received multiple units of packed red blood cells (PRBC's). You would anticipate a 2-3DPG change that would cause the oxyhemoglobin dissociation curve to shift;
   a. right
   b. left
   c. up
   d. down

10. Your patient has a probable closed head injury and has been managed by BLS first responders for almost an hour prior to your arrival. They have inserted a CombiTube and ventilated at a rate of 24 breaths per minute with high concentration O_2 throughout their care. Assuming this is the only injury present, you would anticipate which of the following?
    a. the SpO_2 will read high as a result of excellent oxygenation and good overall cerebral oxygenation is very likely
    b. the SpO_2 will read high complicated by a left shift in the oxyhemoglobin dissociation curve, good cerebral oxygenation is questionable
    c. the SpO_2 will read low complicated by a right shift in the oxyhemoglobin dissociation curve, good cerebral oxygenation is questionable
    d. the SpO_2 will read high as will the $EtCO_2$ due to the poor gas exchange that occurs when using a CombiTube

11. Your patient's labs state the current lactate level is 4.2mmol/L. That value suggests;
    a. nothing, the value is normal
    b. aerobic metabolism is adequate systemically
    c. anaerobic metabolism is adequate systemically
    d. anaerobic metabolism is occurring

12. Elimination of fixed acids will occur via;
    a. the respiratory system primarily
    b. the renal system only
    c. the bicarbonate buffering system
    d. the respiratory, renal and bicarbonate buffering systems

13. Increased CO_2 removal is facilitated by;
    a. increasing FiO_2
    b. Bohr effect
    c. Haldane effect
    d. Boyle's Law

14. The majority of CO_2 is transported;
    a. bound to albumin in the plasma
    b. bound to hemoglobin as a carbamino group
    c. as bicarbonate in plasma
    d. as CO_2 in the plasma

15. The relationship of $EtCO_2$ to $PaCO_2$ should;
    a. be exactly the same
    b. demonstrate $EtCO_2$ is slightly lower than $PaCO_2$
    c. demonstrate $PaCO_2$ is slightly lower than $EtCO_2$
    d. there is no relationship between these values because they are measured differently

16. As $PaCO_2$ decreases by 10mmHg, you would anticipate which of the following?
    a. K^+ increase of 0.5mEq/L and pH increase of 0.8
    b. K^+ decrease of 0.5mEq/L and pH increase of 0.8
    c. K^+ increase of 0.5mEq/L and pH decrease of 0.8
    d. K^+ decrease of 0.5mEq/L and pH decrease of 0.8

17. Your patients ABG's report a current pH of 7.52. Previously the pH was 7.41 and K^+ was 4.7mEq/L. You would anticipate the current K^+ to be approximately;
    a. 5.3mEq/L
    b. 5.0mEq/L
    c. 4.1mEq/L
    d. 3.5mEq/L

18. Your patient's initial ABG values are pH 7.36, $PaCO_2$ 51torr, PaO_2 104torr, HCO_3 27, SaO_2 97%, $EtCO_2$ 44torr. Your ventilator has been managing the patient in transport for 45 minutes and your $EtCO_2$ is now showing 35. All other hemodynamic parameters are without change from your initial assessment. You would anticipate which of the following;
    a. pH 7.34
    b. PaO_2 94torr
    c. SaO_2 100%
    d. $PaCO_2$ 42torr

19. Your patient's ABG values are; pH 7.29, $PaCO_2$ 18, PaO_2 110, HCO_3 17. With these ABG's you would expect to find which of the following;
    a. bradypnea
    b. hyperkalemia
    c. hyponatremia
    d. hyperchloremic alkalosis

20. When treating life threatening hyperkalemia, primary focus rests with;
    a. fluid resuscitation
    b. furosemide(*Lasix*®) and/or sodium polystyrene sulfonate (*Kayexalate*®)administration
    c. albuterol (*Ventolin*®), dextrose and insulin administration
    d. sodium bicarbonate and calcium chloride administration

21. Potassium administration;
    a. may be given via a peripheral IV
    b. should not exceed 0.5-1.0mEq/kg/hr
    c. typically occurs at 10-20mEq/hr
    d. all of the above

22. Your patient's ABG's are as follows; pH 7.48, $PaCO_2$ 30, PaO_2 52, HCO_3 22. This should be interpreted as;
    a. uncompensated respiratory alkalosis
    b. compensated respiratory alkalosis
    c. uncompensated metabolic alkalosis
    d. none of the above

23. Your patient's ABG's are as follows; pH 7.32, $PaCO_2$ 24, PaO_2 62, HCO_3 12. This should be interpreted as;
    a. uncompensated metabolic acidosis
    b. compensated metabolic acidosis
    c. partially compensated respiratory acidosis
    d. partially compensated metabolic acidosis

24. Your patient's ABG's are as follows; pH 7.36, $PaCO_2$ 28, PaO_2 92, HCO_3 19. This should be interpreted as;
    a. normal
    b. compensated respiratory acidosis
    c. partially compensated metabolic acidosis
    d. compensated metabolic acidosis

25. Your patient's ABG's are as follows; pH 7.22, $PaCO_2$ 68, PaO_2 56, HCO_3 18. This should be interpreted as;
    a. uncompensated respiratory acidosis
    b. partially compensated respiratory acidosis
    c. uncompensated metabolic acidosis
    d. mixed disturbance

26. Your patient's ABG's are as follows; pH 7.55, $PaCO_2$ 62, PaO_2 212, HCO_3 45. This should be interpreted as;
    a. uncompensated respiratory alkalosis
    b. partially compensated metabolic alkalosis
    c. uncompensated metabolic alkalosis
    d. compensated metabolic alkalosis

27. Your patient's ABG's are as follows; pH 7.48, $PaCO_2$ 25, PaO_2 56, HCO_3 24. Which of the following would be the most likely cause?
    a. benzodiazepine overdose
    b. chronic alcohol abuse
    c. diabetic ketoacidosis
    d. steam inhalation injury

28. Your patient's ABG's are as follows; pH 7.22, $PaCO_2$ 17, PaO_2 88, HCO_3 11. Which of the following would be the most likely cause?
    a. benzodiazepine overdose
    b. chronic alcohol abuse
    c. diabetic ketoacidosis
    d. steam inhalation injury

29. Your patient's ABG's are as follows; pH 7.24, $PaCO_2$ 60, PaO_2 68, HCO_3 23. Which of the following would be the most likely cause?
    a. benzodiazepine overdose
    b. chronic alcohol abuse
    c. diabetic ketoacidosis
    d. steam inhalation injury

30. Your patient has been diagnosed with an acute myocardial infarction complicated by pulmonary edema. Onset of chest pain was 45 minutes ago. You would anticipate his ABG's to demonstrate;
    a. uncompensated metabolic acidosis
    b. uncompensated respiratory alkalosis
    c. compensated respiratory acidosis
    d. partially compensated respiratory acidosis

31. Which of the following will be most effective in treating respiratory acidosis?
    a. bicarbonate administration
    b. addition of PEEP
    c. maximizing FiO_2
    d. increasing minute ventilation (Ve)

32. Which of the following will be most effective in treating hypoxia?
    a. maximizing FiO_2
    b. addition of PEEP
    c. increasing minute volume
    d. increasing plateau pressure

33. Which of the following is most commonly related to barotrauma from ventilators?
    a. high levels of PEEP
    b. high levels of FiO_2
    c. high peak inspiratory pressures
    d. high plateau pressures

34. Hyperthermia and early acetylsalicylate poisoning would be most likely to cause;
    a. respiratory acidosis
    b. respiratory alkalosis
    c. metabolic acidosis
    d. metabolic alkalosis

35. Opioid over-administration would most likely cause;
    a. respiratory acidosis
    b. respiratory alkalosis
    c. metabolic acidosis
    d. metabolic alkalosis

36. Renal failure would be most likely to cause;
    a. respiratory acidosis
    b. respiratory alkalosis
    c. metabolic acidosis
    d. metabolic alkalosis

37. Continuous OG/NG suctioning would be most likely to cause;
    a. respiratory acidosis
    b. respiratory alkalosis
    c. metabolic acidosis
    d. metabolic alkalosis

38. To increase PaO_2 you should increase __________ first.
    a. PEEP
    b. tidal volume
    c. respiratory rate
    d. FiO_2

39. To assure optimal O_2 delivery to the pulmonary capillaries you must have;
    a. adequate tidal volume
    b. adequate alveolar volume
    c. adequate peak inspiratory pressure
    d. adequate hemoglobin

40. Volume targeted ventilators;
    a. terminate the ventilation based on airway pressure
    b. terminate the ventilation based on volume delivered
    c. terminate the ventilation based on the mode setting (i.e.- SIMV vs. AC)
    d. none of the above

41. A disadvantage of pressure-limit ventilation is;
    a. poorly sedated patients may be hypoventilated
    b. barotrauma routinely results
    c. PEEP cannot be employed with this ventilation technique
    d. high FiO_2's are required for adequate oxygenation

42. Patients ventilated in assist control mode;
    a. cannot take a spontaneous breath
    b. are not assisted when taking spontaneous breaths
    c. are at risk for 'stacking' with subsequent barotrauma
    d. are typically under ventilated

43. SIMV mode is commonly used because;
    a. spontaneous breaths by the patient are prohibited and finer control of ventilation occurs
    b. PEEP is always employed and thus better oxygenation occurs
    c. smaller tidal volumes are utilized minimizing barotrauma
    d. synchronization of the ventilator with spontaneous breaths is optimized

44. Barotrauma caused by ventilators is best prevented by monitoring;
    a. PIP
    b. PEEP
    c. P_{plat}
    d. P_{AW}

45. Which of the following is true concerning ventilation?
    a. high flow rates will provide more laminar flow in large airways
    b. laminar flow is optimized by optimizing the use of small airways with volume
    c. laminar flow is best in the ventilator circuit and endotracheal tube
    d. resistance to flow is maximized in the ventilator circuit and large airways

46. Normal minute volume should be;
    a. 2-6L/min
    b. 8-12L/min
    c. 6-10ml/kg
    d. 4-8L/min

47. Which of the following would be least likely to cause a high pressure alarm?
    a. mucous plug
    b. pneumothorax
    c. ETT dislodgment to the glottis/esophagus
    d. inadequate sedation of the ventilated patient

48. Which of the following would be most likely to cause a low pressure alarm?
    a. circuit disconnect
    b. low oxygen supply pressure
    c. punctured ETT cuff
    d. all of the above are very likely

49. Your patient is demonstrating a sudden elevated PIP with normal/unchanged P_{plat}. The most likely cause would be;
    a. pneumothorax
    b. ARDS
    c. pulmonary edema
    d. asthma

50. Your patient is demonstrating a sudden elevated PIP and P_{plat}. Your evaluation should focus initially on;
    a. performing a 12-lead ECG to identify a probable AMI with secondary pulmonary edema
    b. adding PEEP to treat the ARDS occurring at this time
    c. administration of bronchodilators with an increase in FiO_2
    d. identification of a pneumothorax with subsequent decompression

51. Your patient was found unconscious/unresponsive in an older home. Vital signs are HR 110, RR 40, BP 102/44, SpO_2 100%, Temp 37° C. ABG's indicate: pH 7.23, $PaCO_2$ 17mmHg, PaO_2 82mmHg, HCO_3 14, SaO_2 68% (FiO_2 1.0). Which of the following should be requested first?
    a. CBC
    b. Standard chemistry with anion gap and osmolar gap
    c. Run the ABG's again with a carboxyhemoglobin (COHgb) and methemoglobin (MetHgb) level
    d. Urine tox screen

52. Immediate care of the patient described above should include all but which of the following;
    a. FiO_2 1.0
    b. increased minute ventilation
    c. identification of the closest hyperbaric chamber
    d. transport by ground if possible, otherwise use a pressurized aircraft

53. Standard adult tidal volume calculation should begin with;
    a. 10-15ml/kg
    b. 6-10ml/kg
    c. 4-10ml/kg
    d. 10-12ml/kg

54. When calculating alveolar minute ventilation you must subtract dead space from the minute ventilation. Dead space is approximated using the formula;
    a. 40% of the tidal volume or approximately 4ml/kg
    b. 33% of the tidal volume or approximately 1ml/lb
    c. 20% of the tidal volume or approximately 1ml/kg
    d. 7.5% of the tidal volume or approximately 150ml

55. Your patient is demonstrating a respiratory acidosis. Your initial therapy consisted of increasing the tidal volume while leaving the rate at 16/min. Your PIP is now 42, your P_{plat} is 31. Your next ventilator adjustment would likely be;
    a. decrease the tidal volume and increase the rate
    b. increase the rate only
    c. add PEEP
    d. add PEEP and increase FiO_2

56. Signs of acute respiratory distress include all but which of the following;
    a. tripod positioning
    b. PaO_2 < 60mm Hg
    c. accessory muscle use
    d. reports of fatigue

57. The acutely deteriorating respiratory patient will commonly exhibit;
    a. pulmonary vasodilation
    b. pericardial halo on radiographic examination
    c. $PaCO_2$ >55mm Hg
    d. epistaxis

58. Your patients respiratory rate has dropped from 24 to 18 breaths per minute and their tidal volume had maintained at 500ml. From the information provided, you can tell they are experiencing a;
    a. drop in minute volume
    b. drop in exhaled tidal volume
    c. drop in PaO_2
    d. drop in level of consciousness

59. All but which of the following will directly promote bronchodilation?
    a. albuterol (*Proventil®*, *Ventolin®*)
    b. ketamine (*Ketalar®*)
    c. terbutaline (*Brethine®*)
    d. methylprednisolone (*Solu-Medrol®*)

60. COPD is a form of ______________ lung disease.
    a. restrictive
    b. obstructive
    c. constrictive
    d. mucogenic

61. COPD patients commonly demonstrate which of the following;
    a. thrombocytopenia
    b. leukocytopenia
    c. polycythemia
    d. erythrocytopenia

62. Radiographic examination of the acute asthma victim may reveal;
    a. sharp costophrenic angles
    b. profound Kerley's B-lines
    c. hyperdensity of lung fields
    d. flattened diaphragm

63. Radiographic examination of the acute pulmonary edema victim may reveal;
    a. sharp costophrenic angles
    b. profound Kerley's B-lines
    c. hypodensity of lung fields
    d. narrow mediastinum

64. Radiographic examination of the ARDS victim may reveal;
    a. "patchy" infiltrates
    b. "ground-glass" appearance
    c. obliterated costophrenic angles
    d. all of the above

65. Your 28 year old female patient has presented with acute onset respiratory distress. She reports, "As a child I had asthma or bronchitis, I'm not sure which." Today she was arguing with her boyfriend when she could no longer catch her breath. Initial studies at the sending facility demonstrate normal electrolytes, an H&H of 13 & 33.2, ABG's show a pH 7.29, PaO_2 55mmHg, $PaCO_2$ 78mmHg, HCO_3 28 and BE -7 on room air at sea level. Other tests reveal a positive D-dimer, urine is negative for ketones, blood or glucose and an HCG (pregnancy test) is positive

as well. The RN taking care of the patient states that she initially had an SpO_2 of 48% and it only improved to 53% with a non-rebreather face mask at 15lpm. She adds, "Her lungs sound bad to me but I can hardly hear anything with this cheap stethoscope and the doctor yelling at me." Based on the information provided, the most likely cause for the respiratory distress would be;

a. acute exacerbation of asthma
b. psychogenic hyperventilation
c. hypoglycemia
d. pulmonary embolism

66. Your patient has been in the referring facility's ICU for almost two weeks with a diagnosis of "urosepsis". Their respiratory status has been deteriorating over the last four days and today the managing internist was heard to dictate, "..patient has bi-basilar patchy infiltrates on today's chest films". He then requested the patient be transferred to a major tertiary care facility. What is a likely cause of the respiratory distress?

a. Excessive WBC proliferation with hypomotility of hemoglobin
b. Adult Respiratory Distress Syndrome (ARDS)
c. Tuberculosis
d. anemia with 2-3DPG deficiency syndrome

67. Your 320 pound patient was initially admitted by the sending facility for a Roux-en-Y procedure. Now, seven hours post-op, the patient has developed restlessness, acute dyspnea and chest pain. 12-lead EKG does not show significant ST changes in any lead but large r-waves are noted in V1 and V2. Vitals are; HR 124, BP 91/44 & RR 32. Heart tones indicate a clearly audible pulmonary S2. You should suspect;

a. spontaneous pneumothorax
b. pulmonary embolism
c. hemorrhagic shock
d. pleural effusion

68. You are called to transport yet another AMI patient from a rural facility to a tertiary care facility. Upon arrival you are told the patient was diagnosed and treated for his AMI four hours prior with good response. However, one hour ago he complained of severe chest pain, shortness of breath and became increasingly restless. Physical exam reveals an audible systolic murmur auscultated best at the apex of the heart and significant pulmonary edema. This change in his status is most likely caused by;
    a. ruptured papillary muscle
    b. cardiogenic shock
    c. Bezold-Jarish syndrome
    d. aortic stenosis

69. You are requested to transport an 81 year old female to a tertiary care facility for admission to a critical care unit. She was seen at the sending facility for dyspnea and altered LOC. An anterior/posterior chest film demonstrates infiltrates in the right middle lobe. Based on this information alone, the most likely cause of the respiratory distress is;
    a. sepsis
    b. pneumonia
    c. emphysema
    d. ARDS

70. The body's ability to respond to changing respiratory needs is influenced the fastest by;
    a. hypoxia
    b. hypercarbia
    c. hydrolysis
    d. hydrogen

71. Your patient was admitted to the sending facility for dyspnea that rapidly progressed to respiratory arrest and subsequent intubation. Upon arrival you note the following $EtCO_2$ waveform (see below right). The most likely cause of the ventilatory failure based on the information and diagram given is;
    a. pneumonia
    b. ARDS
    c. pulmonary embolism
    d. asthma

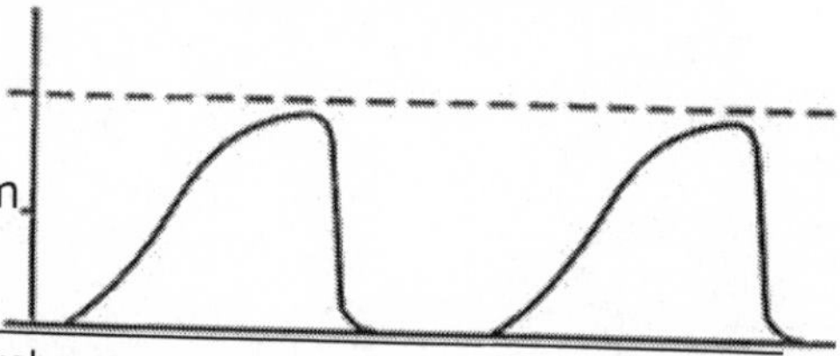

72. You are transporting an intubated patient. While monitoring in flight you note a high pressure alarm on your ventilator and the following $EtCO_2$ waveform (see below). The first course of action should be;
    a. re-paralyze the patient
    b. re-sedate the patient
    c. manually ventilate the patient
    d. no immediate action is necessary

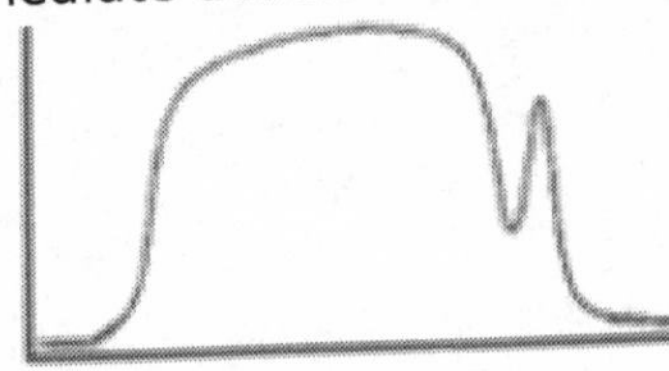

73. The most likely cause of the $EtCO_2$ waveform seen in the previous question would be;
    a. inadequate tidal volume
    b. inadequate sedation
    c. hiccups
    d. ventilator failure

74. Your intubated & mechanically ventilated patient's $EtCO_2$ waveform baseline is slowly climbing with a current level of 8mm. Which of the following is the most likely cause?
    a. increased CO_2 production
    b. partially obstructed exhalation valve
    c. decreased spontaneous respiratory effort
    d. increased minute ventilation

75. Transport monitor pulse oximetry utilizes infrared technology to measure;
    a. oxygen saturation
    b. oxyhemoglobin saturation
    c. plasma oxygen partial pressure tension
    d. carboxyhemoglobin saturation

KEY & RATIONALE

RESPIRATORY DISEASE, ACID-BASE DISTURBANCES & VENTILATOR MANAGEMENT

1. D The bicarbonate buffer system is an almost instantaneous system that manages acid-base derangement. While this system is extremely fast it is also easily overwhelmed. The next fastest mechanism would be via the respiratory system followed by the renal and lastly, the phosphate buffering systems.

2. D pH is 'per Hydrogen'. You undoubtedly knew pH referred to the acid-base status and may have selected carbon dioxide. In my experience, I have seen most of Paramedic and RN medical education emphasize that carbon dioxide is "an acid". While carbon dioxide is critical in the management of acid, it is actually the exchange of Hydrogen that causes the chemical properties of an acid. You need to adopt a new paradigm. "Acid" is hydrogen. Managing acid-base disturbances is all about managing hydrogen molecules.

TEST TIP

Hydrogen (H^+) ⇒ Acid

3. D During aerobic metabolism cells incorporate primarily glucose and oxygen to produce carbon dioxide, water and adenosine triphosphate (ATP). Obviously there is much more to the specific reaction, the Kreb cycle, the electron transport chain and the inner workings of the mitochondria, but this basic concept is key.

$$H^+ + HCO_3^- \Leftrightarrow H_2CO_3 \Leftrightarrow CO_2 + H_2O$$

4. D As stated in the previous question's rationale, normal aerobic metabolism produces CO_2, water and ATP. ATP is the purpose of the reaction or goal product. The byproducts would be CO_2 and water. Lactate and pyruvate can be involved in metabolism but these are not the normal <u>end</u> byproducts of *aerobic* metabolism.

5. A The Bohr effect is the primary mechanism, which prompts the 'unloading' of oxygen at the tissue level. As stated in the answer, the binding of CO_2 with hemoglobin causes a 'conformation change' or change in shape, which decreases the molecule's affinity (read 'desire') to hold oxygen. The Bohr effect is associated with a right shift in the oxyhemoglobin dissociation curve and as such answer B is incorrect. The reverse-Bohr effect occurs in the lungs as CO_2 is released from the hemoglobin causing a return in conformation (shape) that again increases the affinity for oxygen to bind. The Bohr effect occurs primarily in the tissues needing oxygen, not the areas providing oxygen. Carbon monoxide (CO) and carboxyhemoglobin (COHgb) actually cause an increased affinity of the Hgb for oxygen with a subsequent left shift of the oxyhemoglobin dissociation curve. As a result, the hemoglobin which are not bound to CO and still have oxygen bound (oxyhemoglobin), have an increased affinity for that oxygen and do not want to release it as well as they did prior to the CO introduction into the system.

TEST TIP
Bohr effect ⇒ release of O_2 ⇒ tissue level

6. C There are many things that could cause or exacerbate the scenario at hand. Be sure not to focus energy there. Of the options presented *they* want you to identify the type of hypoxia involved. Hypoxic hypoxia would suggest not enough oxygen is presented to the lungs. If the SaO_2 is 98% he's clearly getting oxygen to the alveoli, it's crossing the plasma and saturating the hemoglobin. Histotoxic hypoxia might be suspected if everything else involved in oxygen carrying capacity were normal otherwise. His Hgb is clearly not 'normal'. Also, from a testing perspective, if histotoxic hypoxia were to be sought, they would undoubtedly need to present more of a history suggestive of an exposure to toxins of some sort. Stagnant hypoxia can be ruled out by the adequate blood pressure, appropriate tachycardia and pale, yet dry skin. He's clearly moving blood; it's just not carrying enough oxygen. Hypemic hypoxia is your only rational choice here.

TEST TIP
"anemic ⇒ hypemic"

7. C While answer B may be enticing; it is placed there for that exact reason. You must dissociate 'hard' values for vital signs from shock. Ultimately shock indicates there is inadequate tissue perfusion. Some tissue is not getting oxygen. As such anaerobic metabolism is in progress in that tissue and the primary product of such is lactic acid. A lactate level of 0.7-2.1 mEq/L is considered 'normal' or for ease of memory remember 'Lactate of 1 to 2'. Typically a lactate >4 is considered clinically significant and the etiology sought out. Remember a positive D-Dimer simply means a clot formed and has begun to stabilize somewhere in the body as the fibrin stabilizing factor (factor XIII) has crosslinked the fibrin filaments of an actual thrombus. While many things can cause a low potassium you should anticipate a high potassium in anyone with an acute shock state as acidosis will mobilize more potassium from intracellular compartments to intravascular compartments.

TEST TIP
shock ⇒ lactate
lactate >4 ⇒ significant
acidosis ⇒ hyperkalemia

8. A Anticipate a **R**ight shift of the oxyhemoglobin dissociation curve anytime you have a **R**aised temperature, **R**aised 2-3DPG, **R**aised acidotic state or a **R**educed oxygenation state. The curve only shifts left and right. The PRBC administration would lower the 2-3DPG via citrate binding and contribute to a **L**eft shift, however this is minimal compared to the **R**ight shift that will be caused by the significantly **R**aised acidosis.

TEST TIP
Oxyhemoglobin dissociation curve
RIGHT shift ⇒ Hgb Releases O_2
RIGHT shift caused by; Raised temp, Raised 2-3DPG, Raised acidosis, Reduced oxygenation

Oxyhemoglobin dissociation curve

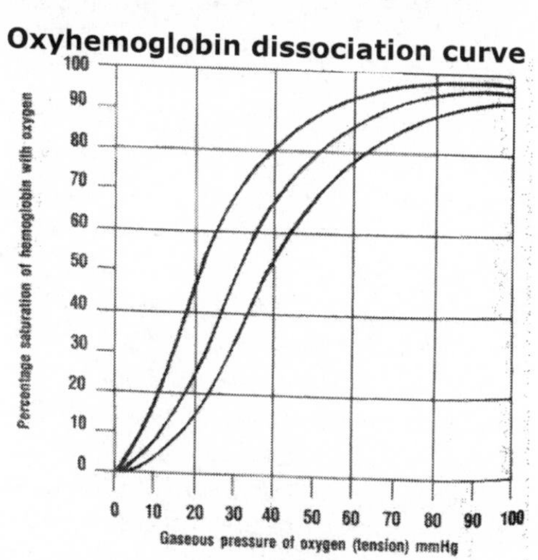

Normal curve in the center (ph 7.4), right and left shifts depicted as well. Left shift curve correlates with pH 7.6. Right shift curve correlates with pH 7.2.

9. B Massive transfusion of PRBC's can cause a lowering of the 2-3DPG. Citrate is added to stored PRBC's as a calcium-chelating agent to prevent spontaneous clotting while in storage. In addition to chelating the calcium it tends to destroy 2-3DPG as well. As such, massive transfusions can overwhelm the patient's calcium stores, specifically if they were previously hypocalcemic and/or their liver function is not optimal. It is commonly reported that the liver can metabolize the amount of citrate in one unit of PRBC's in approximately 5 minutes, so unless you're administering blood faster than one unit every 5 minutes, citrate toxicity is a moot point. This statement assumes you have good liver function and more importantly good hepatic flow. Patients receiving massive transfusions, specifically in the prehospital environment are likely deficient in one or both areas, and giving a unit of blood in under 5 minutes is not impossible or unheard of, especially with a motivated flight team and an abundant supply of 14ga angiocaths. When possible, guide calcium administration by current lab values, remember no treatment is truly benign. A **L**eft shift in the oxyhemoglobin dissociation curve is typically caused by a**L**kalosis, **L**ow 2-3DPG, **L**ow temperature or **L**ots of carbon monoxide. Remember a left shift is indicative of a reverse-Bohr effect so we see the hemoglobin's affinity for oxygen increase (ho**L**d oxygen) and as such SpO_2 may be very high but off-loading to the tissues is actually impaired.

TEST TIP
Oxyhemoglobin dissociation curve
LEFT shift ⇒ Hgb hoLds O_2
LEFT shift caused by; Low temp, Low 2-3DPG, aLkalosis (Low acidosis), Lots of CO

10. B This scenario builds on the concepts of the previous question's rationale. Hyperventilation will cause a respiratory alkalosis. Alkalosis will cause a left shift of the oxyhemoglobin dissociation curve and as such the hemoglobin's affinity for oxygen climbs. SaO_2 may be high but it may not be indicative of tissue oxygenation. Continuous non-discriminate hyperventilation is no longer endorsed because the vasoconstriction caused by such alkalosis is not specific to the injury site and causes secondary hypoxic brain injury to previously healthy regions.

11. D A normal lactate is 0.7-1.2mmol/L. Greater than 4mmol/L is considered clinically significant for shock.

12. B "Fixed" acids are acids that cannot be manipulated by the respiratory system. They may be manipulated by the bicarbonate buffer system but they are not *eliminated* via that route. An example of a fixed acid would be aspirin.

13. C The Haldane effect is what enables us to remove exponentially large amounts of CO_2 relative to the pressure gradient presented to the alveolar interface. Increasing FiO_2 will not help, the Bohr effect specifically addresses oxygen loading/unloading. You should have identified the Boyle's Law answer as a 'left field answer' and immediately eliminated it from your options.

14. C Carbon dioxide is transported by every means listed there, however the overwhelming majority is as bicarbonate (HCO_3).

TEST TIP
Vast majority of CO_2 is transported in the plasma as HCO_3^-

15. B $EtCO_2$ and $PaCO_2$ have a direct relationship. The $EtCO_2$ can never be higher than the $PaCO_2$, if it were, the pressure gradient would dictate that CO_2 move from the lungs *into* the plasma. Even in an apneic patient they never *completely* equalize due to a minimal diffusion block.

16. B This relationship is well established and key to estimating your effects on pH and K^+ as you manipulate the $PaCO_2$. Keep in mind that as long as hemodynamics are consistent, changes made to $EtCO_2$ should reflect almost identically in the $PaCO_2$. Note I said the change is identical, not the value. I.e.- your $EtCO_2$ is 45 and $PaCO_2$ is 50. If you drop the $EtCO_2$ to 35 you can anticipate you changed the $PaCO_2$ to 40, they will both drop 10mm Hg.

TEST TIP

$$\Delta PaCO_2 \text{ 10mm Hg} \Rightarrow \Delta pH\ 0.08 \Uparrow\Downarrow$$

$$\Delta pH\ 0.1 \Rightarrow \Delta K^+\ 0.6 \Uparrow\Downarrow$$

$$\Delta PaCO_2 \text{ 10mm Hg} \propto \Delta K^+\ 0.5$$

$$\Delta EtCO_2 \propto \Delta PaCO_2$$

17. C Using the relationship mentioned in the question above will get you very close. Another clinical mathematical relationship to consider is that for every change in pH of 0.1 you will see a change in K^+ of 0.6 the opposite direction.

18. D Assuming there is no change in hemodynamic status (as stated), the changes in $EtCO_2$ will be reflected in the $PaCO_2$. Dropping the $PaCO_2$ would increase the pH, not drop it. Dropping the $PaCO_2$ by 9 won't have any appreciable effect on PaO_2 or SaO_2.

19. B This patient is in a metabolic acidosis, which is causing a partially compensating hyperventilation. The drop in pH suggests that serum potassium should be elevated. You may see what appears to be a normal, or low potassium in a patient presenting in a metabolic acidosis such as diabetic ketoacidosis. Why is the potassium not high with DKA? Right! They've been diuresing the potassium off as part of their disease process. This is the very reason we don't immediately correct metabolic acidosis in a DKA patient with something like sodium bicarbonate. A rapid shift to a normal pH (from acidosis) would precipitate a tremendous fall in the potassium level ($pH\Uparrow \Rightarrow K^+\Downarrow$) from the level at which they presented and initially appeared "normal".

20. D All of the therapy options listed may be appropriate but the "primary focus" is on altering cellular resting potentials and action potential thresholds. Sodium bicarbonate will raise the pH and force the potassium intracellular (thereby lowering the resting potential), effectively 'hiding' it from the heart by getting out of circulation. Meanwhile calcium chloride will elevate the action potential threshold away from the resting potential thereby making it harder for an action potential to meet threshold and cause an action potential to occur that could result in a dysrrhythmia.

21. D Option B is the most extreme administration limit and would only be used via a central venous access with close monitoring. Peripheral administration is commonly done, typically at 10mEq/hr. If administration at 20mEq/hr is desired, administration via central venous access is typically recommended. Safe administration can be performed up to 1mEq/kg/hr per some reputable texts.

22. A The pH tells you it must be uncompensated or partially compensated. The pH also tells you it's alkalosis. The $PaCO_2$ is low suggesting alkalosis while the HCO_3 is normal. There is a respiratory etiology without any renal compensation.

23. D The pH tells you it must be uncompensated or partially compensated. The pH also tells you it's acidosis. The $PaCO_2$ is low suggesting alkalosis while the HCO_3 is low suggesting acidosis. Therefore there is a metabolic etiology without *adequate* respiratory compensation.

24. D The pH tells you it must be compensated or normal. The pH also tells you it's on the acidosis side of a perfect 7.4. The $PaCO_2$ is low suggesting alkalosis while the HCO_3 is showing acidosis. A metabolic etiology exists with respiratory compensation back to within normal range, hence 'compensated'.

25. D The pH tells you it must be uncompensated or partially compensated. The pH also tells you it's acidosis. The $PaCO_2$ is high suggesting acidosis while the HCO_3 is also low, suggesting acidosis. Therefore both metabolic and respiratory systems are contributing to the acidotic state such as in a post-arrest patient. This is commonly called a "mixed gas", "mixed acidosis" or "mixed disturbance".

26. B The pH tells you it must be uncompensated or partially compensated. The pH also tells you it's alkalosis. The $PaCO_2$ is high suggesting acidosis while the HCO_3 is very high suggesting alkalosis. This shows a metabolic etiology without *adequate* respiratory compensation.

27. D The ABG's suggest an uncompensated respiratory alkalosis quite likely caused by hyperventilation in response to hypoxia. While a benzodiazepine overdose can cause hypoxia it would be secondary to *hypo*ventilation, clearly not occurring here. DKA will cause hyperventilation but it's etiology rests in an underlying metabolic acidosis, which is not occurring here either. The steam inhalation injury has caused edema of airways and hypoxia with resultant hyperventilation as an attempt to compensate. Recall that CO_2 transfers much better than oxygen through tissues and water.

28. C This gas is typical of DKA. You will find metabolic acidosis with secondary hyperventilation in an attempt to compensate.

29. A This gas indicates the patient is retaining CO_2 secondary to inadequate minute ventilation. This gas also suggests an acute state, as the HCO_3 has not left normal range to attempt to compensate yet.

30. B This patient is likely to be hypoxic secondary to the pulmonary edema. His first compensatory action will be to hyperventilate. Off-loading of CO_2 will increase pH showing a respiratory alkalosis. Due to the acute timing of the situation, the renal system will not have had time to alter the HCO_3 levels to compensate for the hyperventilation.

31. D Respiratory acidosis is secondary to retained CO_2. Retained CO_2 is a function of inadequate alveolar minute ventilation. Bicarbonate will alter the pH temporarily but without the ability to increase CO_2 off-loading, the system will return to it's previous state very quickly. PEEP and FiO_2 have no beneficial influence on CO_2 elimination, with the exception that PEEP may assist in holding de-recruited alveoli open in the lung of a COPD/ARDS victim. Open alveoli ⇒ improved gas exchange. The "normal" lung doesn't typically need this and as such, PEEP is not a typical management tool for respiratory acidosis.

TEST TIP

To *decrease* $PaCO_2$

Increase Ve

1. **Verify maximum safe Vt (keep P_{plat}<30-35)**
2. **Increase F (rate)**

To *increase* $PaCO_2$

Decrease Ve

1. **Decrease frequency (rate)**
2. **Decrease Vt no more than absolutely necessary (prevent de-recruitment)**

32. A Maximizing FiO_2 will increase PaO_2 more respective to PEEP. Once FiO_2 has been maximized, *then* addition of PEEP is considered most beneficial. Increasing minute volume can improve O_2 loading but very minimally.
(Recall the alveolar air equation: $PAO_2 = FiO_2(P_B - P_W) - 1.2(PaCO_2)$, decreasing $PaCO_2$ will increase PAO_2 but only at a 1.2:1 ratio respectively). Increasing plateau pressure will only make your patient more likely to suffer barotrauma.

TEST TIP

To increase PaO_2;

1. **Verify normal/adequate Ve**
2. **Maximize FiO_2**
3. **Add PEEP**
4. **Consider reducing I:E ratio**
5. **Invert I:E ratio**

33. D High levels of PEEP can contribute to high P_{plat}'s if not properly managed but one may administer a very high PEEP and still maintain a normal P_{plat}. High FiO_2 levels are associated with oxidative radicals but a high percentage of oxygen in and of itself does not cause barotrauma. Peak inspiratory pressures are a measurement of airway resistance and in the normal patient should reflect the plateau pressures relatively closely, again however, this is not without exception. The asthmatic patient may exhibit very high PIP's while having normal P_{plat}'s. Thus the PIP in and of itself doesn't cause the barotrauma, barotrauma is primarily a result of high alveolar pressures, which are best assessed by plateau pressure (P_{plat}).

34. B Hyperthermia and acetylsalicylate poisoning both promote hyperventilation. The first due to hypoxia from the hypermetabolic state, the latter causes a metabolic acidosis but initially promotes a direct thalamic hyperactivity in the respiratory center.

TEST TIP
Aspirin promotes hyperventilation directly

35. A Opioid administration commonly causes respiratory depression with a decrease in minute ventilation. As such respiratory acidosis would occur as CO_2 is retained.

36. C The renal system is responsible for eliminating all fixed-acids as well as the final compensations that the respiratory system cannot make. In addition to the acid management, the renal system is responsible for retention and manufacturing of new bicarbonate. With renal failure there is a net failure to eliminate acid and retain, or make new bicarbonate, thus, metabolic acidosis.

37. D "Continuous" suctioning of the stomach causes metabolic alkalosis via multiple mechanisms. With the removal of stomach acid this causes an obvious depletion of hydrochloric acid and stimulates the stomach to make more. The process of making more stomach acid in turn generates bicarbonate via a carbonic anhydrase reaction, which normally would be utilized at a later point during digestion to neutralize the gastric contents from their normal pH of 2-3 to a more intestine-friendly pH. Since the acid never makes it to the intestines because of its removal via the NG/OGT, the excess bicarbonate is never 'used' in the gut and is therefore synergistic with the alkolotic condition created by the acid removed initially.

38. D This question is a little tricky or misleading. I think it's fair to assume the patient is being ventilated or breathing spontaneously at an 'adequate' alveolar minute volume. If the patient were breathing very shallow, or not at all, tidal volume and rate would be addressed first obviously. However, on the assumption that ventilation is taking place at a normal depth and rate, FiO_2 will increase oxygenation much faster than PEEP. Admittedly, once FiO_2 has been maximized, PEEP is your next best choice but FiO_2 is "King". Refer to the alveolar air equation, $PAO_2=FiO_2(P_B-P_W)-1.2(PaCO_2)$, and you see that to increase the partial pressure of oxygen in the alveoli (PAO_2), the driving force behind moving oxygen into the plasma you must either; increase FiO_2, increase barometric pressure (P_B), reduce the water vapor pressure (P_W) or reduce the $PaCO_2$. You can do almost all of these. Obviously you can increase the FiO_2 from 0.21 to 1.0, almost a 500% increase that is mathematically directly related to PAO_2. You can increase the barometric pressure with PEEP from 760torr (at sea level) to a whopping 767.35torr with 10cm PEEP, slightly less than a 1% mathematical increase, or you can use your handy hyperbaric chamber. PEEP isn't effective at all until you consider it's applied continuously; even so, adding 10cm of PEEP won't be as effective as maximizing FiO_2. To finish the thought process here, you could attempt to reduce the water vapor pressure by reducing the amount of humidity in the inhaled mixture, unfortunately this would require you limit how much humidity the patient donates in the process and this would require you placing your patient into an un-survivable shock state, so we'll just throw that one out. You could hyperventilate their $PaCO_2$ down, this is in fact how mountain climbers are able to maintain survivable oxygenation at very high altitudes, again this is counter-productive in the sick and injured, reeking havoc with acid-base balance, oxyhemoglobin dissociation curves and so on.

39. B To deliver oxygen to the pulmonary capillaries you must first deliver the oxygen to the alveolar-capillary interface. This requires a tidal volume that fills all the alveoli possible (alveolar volume) after filling the dead space. Peak inspiratory pressure is a reflection of airway resistance and doesn't give you accurate assessment of alveolar filling per se. Hemoglobin is required for proper oxygenation of the patient but this question only addresses getting the oxygen to the pulmonary capillaries, hemoglobin would be physiologically 'downstream'.

40. B "Volume-targeted" (aka- "volume-limited") refers to using the volume to trigger the *termination* point when ventilating. Option A would describe pressure targeted or pressure-limited ventilation. The modes, AC, SIMV, etc. all determine when to *initiate* the ventilation, not termination.

41. A "Pressure-limited" (aka- "pressure-targeted") ventilation uses a provider determined pressure limit to trigger termination of the ventilation. As such a poorly sedated or bucking patient will cause spikes in the airway or peak inspiratory pressure, thus triggering the termination point set on the ventilator. Pressure limited ventilation by it's very mechanism is designed to limit or prevent barotrauma. PEEP can and (specifically in neonates and pediatrics), is frequently employed with pressure-limited ventilation. FiO_2 is not a factor in the vent limiting mechanism.

42. C The assist control mode is designed such that the patient receives the fixed 'control' rate and is assisted with another 'full breath' every time the ventilator detects a spontaneous breath via a negative pressure in the airway circuit. As such, slight hiccups or small inhalations for any reason can trigger an additional full breath per the ventilator's current settings. The frequency of the assistance is dependent upon the ventilators sensitivity setting. With this in mind it's easy to see how a patient that hiccups just before a 'scheduled' breath from the ventilator would receive a full breath assist and before exhalation can be completed, receive the next scheduled full breath, thus over-inflating the lungs. Assist control is very effective for the patient hyperventilating in a compensatory manner, (i.e.- DKA). The patient with erratic, non-compensatory breathing patterns are at increased risk of barotrauma when ventilated in AC mode.

43. D SIMV or synchronized intermittent mandatory ventilation attempts to optimize allowing the patient to assist their own minute volume while synchronizing with the desired ventilation parameters.

44. C P_{plat} or plateau pressure is a representation of the static pressure in the pulmonary system. Because this is a static measurement, it is considered more predictive of actual alveolar pressures and is thus best for monitoring to prevent barotrauma. PIP is a reflection of airway resistance and in the patient with normal airways, closely *approximates* P_{plat}. PEEP obviously monitors the minimal pressure maintained during exhalation and P_{AW} is simply a real-time reading of <u>a</u>ir<u>w</u>ay pressure.

45. B Laminar flow is optimal in the small airways as a cumulative unit. Using Poiseuille's Law we can determine that large airways, the circuit and ETT are prone to disorganized, turbulent flow that is only made worse by higher flow rates. By optimizing volume and using <u>all</u> the small airways, the cumulative small airway cross-sections promote flow in a laminar fashion. Airway resistance while factorial in laminar flow is actually maximized with the smaller the airway, not in large airways or the vent circuit.

46. D If you chose C you're brain is tired and you're looking for familiar numbers without really evaluating the question. C would be an accurate tidal volume, not minute volume. It may be time to put the book down and rest the brain. Now if you have trouble recalling the numbers, consider how to calculate minute volume (Ve = Vt X F). Now plug in some standard values,
i.e.- Ve = ((6 to 10ml/kg X 70kg) X 12) = ((420 to 700ml) X 12)
Ve = 5040ml to 8400ml
5.4L to 8.4L/min appears to be the range you might see with a 70kg adult breathing 12 breaths per minute. This would definitely get you 'in the ballpark' to make an educated guess.

TEST TIP
Normal Ve = 4-8L/min
Proper VQ matching suggests Ve should match CO so guess what...
Normal CO = 4-8L/min

47. C Be cautious with questions worded such as this. Which is "least likely" again can be very tricky if placed late in the subject matter and thus when you're fatigued. Obviously pulling the tube into the glottis or the esophagus would most likely give you a low-pressure alarm as it attempts to fill the gut with a ventilation, all the others will promote high pressure alarms.

48. D Don't forget that the low pressure alarm on most ventilators also identifies low O_2 supply pressures, not just low airway pressures.

TEST TIP
Low pressure alarm ⇒ disconnect & supply

49. D A climbing PIP signifies high *airway* pressures. This can be caused by any of the options given. The key here is the constant P_{plat}. The *steady* pressure of the lung parenchyma is suggesting that the lung tissue is normal, thus the problem lies between the ventilator and the lung tissue or alveoli. This differential should lead you to 'tight airways'.

TEST TIP
High PIP ⇒ check for tight 'airways'
High P_{plat} ⇒ causing barotrauma, check volumes

50. D The "sudden" development of an elevated PIP and P_{plat} should tell you the lungs are collapsing. Pressures everywhere are climbing. Identification and treatment of a pneumothorax is paramount, once ruled out then look for other causes of sudden decrease in pulmonary compliance, (flash edema, negative pressure edema, diaphragmatic rupture, etc.) One might compare this to a pericardial tamponade and the hemodynamics seen in that situation. Compare the PIP as similar to the CVP and P_{plat} similar to PCWP. As the numbers rise and become identical that suggests the entire lung apparatus is becoming pressurized from an external mechanism. The pneumothorax acts on the lung in much the same way a pericardial tamponade works on the heart in this respect.

51. C Probably a pretty obvious question but I chose it to illustrate a few key points. This patient is likely suffering from carbon monoxide (CO) poisoning. As such the initial presentation suggests hypoxia but the SpO_2 shows 100%. Remember CO will cause a false high on *pulse* oximeters (SpO_2). The ABG's are key in that you see the SaO_2, a co-oximeter derived saturation is very low while the PaO_2 is pretty good considering. By running the gases again with a carboxyhemoglobin (COHgb) level you will be able to ascertain the degree of poisoning. Normal individuals typically have less than 3% COHgb with smokers and people living in heavily polluted regions showing up to ~10% COHgb. With the SaO_2 sitting at 68% you can predict the COHgb is likely to be around 32%, which would be a very dangerous poisoning.

52. B Increasing minute ventilation will not significantly alter the COHgb levels. The other three options are much more effective in treating carbon monoxide poisoning.

53. B 6-10ml/kg is the current standard to begin your tidal volume calculation. Older standards used the 10-15ml/kg while current pediatric texts will recommend 10-12ml/kg. New research on ARDS is recommending that 4-10ml/kg may be more beneficial but this has not been accepted as the standard in the exam reference texts at the time of this writing.

TEST TIP
Vt = 6 to 10ml/kg

54. B Dead space can be estimated by 1ml/lb of ideal body weight or ~33% of the tidal volume. Subtracting this dead space from your tidal volume reveals the alveolar volume. Calculating alveolar minute volume will reveal why tidal volume changes are more effective at eliminating carbon dioxide in respiratory acidosis than rate changes.

55. B Your initial changes took the pressures up, the most important of which, the P_{plat} is showing 31. Current recommendations suggest keeping P_{plat} below 30-35 if possible. Adding volume is not indicated, neither is reducing it. At this point the best manipulation of minute ventilation would be to simply increase the rate. Volume has been maximized/optimized (you're using the entire lung), use rate to manipulate the carbon dioxide at this point.

56. D Recall your definition of 'signs' vs. 'symptoms'. All of these options are common with acute respiratory distress but 'reports of fatigue' would be a *symptom*.

57. C The pulmonary bed vasoconstricts in the presence of hypoxia as opposed to other system's vasodilation, this is referred to as the Hypoxic Pulmonary Vasoconstrictive Response (HPVR). Pericardial halo is typical of pericardial tamponade, which can result in acute respiratory distress, but this condition is much less common than option C. Epistaxis is simply a distracter.

58. A Minute volume (Ve) is calculated by multiplying tidal volume (Vt) times rate or frequency (F); Ve=Vt X F. Exhaled tidal volume must be measured so from the information provided you cannot definitively state B is true. A reduction in PaO_2 may occur from any number of causes including an acute drop in Ve but the patient who is recovering from a 400 meter dash, for example, should also drop his Ve while the PaO_2 remains constant or even improves. A drop in level of consciousness could lead to a drop in Ve but you have no supporting evidence for that option in the question provided.

TEST TIP
Ve = Vt X F

59. D albuterol, ketamine and terbutaline all provide direct stimulus that promotes bronchodilation. Methylprednisolone (Solu-Medrol®), a steroid, provides for bronchodilation via multiple second-messenger mechanisms and as such, is an indirect bronchodilator.

60. B What does COPD stand for? Restrictive lung disease would include disease states such as pulmonary fibrosis. Constrictive and mucogenic are distracters.

61. C COPD patients will develop polycythemia (high red blood cell quantity) as a compensatory mechanism. This proliferation of red blood cells is what gives COPD'ers the pinkish coloring of their skin and thus the nickname "pink puffers". The same stimulus that promotes polycythemia also promotes elevated 2-3DPG production. If you recall, 2-3DPG acts like a 'crowbar' that is used to pry oxygen off of hemoglobin. So your bonus question here is, "Which way does the oxyhemoglobin dissociation curve shift with a COPD patient?" **R**ight. **R**aised 2-3DPG ⇒ **R**ight shift ⇒ **R**elease oxygen

62. D Asthma patients will demonstrate a narrowed mediastinum, hypodense lung fields, blunted or squared costophrenic angles and a flattened diaphragm, all signs of a hyperinflated chest.

63. B Pulmonary edema patients commonly have fluid deposit to the gravity dependent costophrenic angles causing a hyperdensity and thus 'obliteration' of the costophrenic junction. Kerley's B-lines are vascular engorgements of the pulmonary vasculature and thus these hyperdensities are visible on the X-ray as horizontal lines radiating out from the mediastinum, primarily in the bases. Chronic pulmonary edema patients will typically demonstrate in addition to the Kerley's B-lines, a wide mediastinum with an enlarged heart.

64. D All of the options here are typical 'hints' offered on exams attempting to describe ARDS on a chest film. "Obliterated costophrenic angles" is also commonly used to describe cardiogenic pulmonary edema.

65. D This scenario presents you with what seems a very incomplete and sketchy picture. This is intentional. I have provided just enough information to make all the options appear plausible but only D is quite likely. Tips to lead you towards pulmonary embolism include;

Pregnancy: all pregnant women are by the very physiologic nature of pregnancy, hypercoagulopathic and commonly 'toss clots'

SpO_2: failing to increase SpO_2 substantially from 48% with high flow O_2 suggests a 'shunt', or oxygenation of unperfused lung

$AaDO_2$: The alveolar-arterial oxygen differential is greater than 20. This is calculated for you by most ABG machines, however in this scenario you must calculate it yourself. First I will demonstrate the textbook version(read "long way"), then I will demonstrate an approximating short-cut. To find $AaDO_2$ first calculate the alveolar partial pressure of oxygen (PAO_2) using the formula;

$$PAO_2=FiO_2(P_B - P_W)-0.8(PaCO_2)$$

P_B is barometric pressure, which on a "perfect day" at sea level is 760torr. P_W is water vapor pressure; use 47torr for 100% saturation, which all inhaled air is. A constant multiplier of 0.8 is used instead of the 1.2 cited in previous question rationales because this scenario utilizes 'low flow oxygen'. High flow scenarios should employ the 1.2 constant previously demonstrated.

$$PAO_2 = 0.21(760\text{torr}-47\text{torr}) - 0.8(78\text{torr})$$
$$PAO_2 = 0.21(713\text{torr}) - 62.4\text{torr}$$
$$PAO_2 = 149.7\text{torr} - 62.4\text{torr}$$
$$PAO_2 = 87.33\text{torr}$$

Next calculate the A-a differential ($AaDO_2$) (aka - A-a gradient) using the formula;

$$AaDO_2 = PAO_2 - PaO_2$$
$$AaDO_2 = 87.33\text{torr} - 55\text{torr}$$
$$AaDO_2 = 32.33\text{torr}$$

Now an approximating short cut is to consider what the patient's PaO_2 *should* be. This can be *estimated* by using the calculation FiO_2 X 5. In this case she was on room air at the time of the initial ABG's so FiO_2 = 21%. 21 X 5 = ~100, her PaO_2 *should* be *about* 100mmHg (105mmHg for you OCD types). Her PaO_2 should be almost equal to her PAO_2 less 5 to 15 considering her age and reported health status. Her calculated PAO_2 from the equation above came out to 87.33mmHg, which is pretty close to the estimated 100 -5 to 15 (95-85mm Hg range). So if you had just done the estimation and come up with 100mmHg less 5 to 15 and compared that to her measured PaO_2 of 55mmHg you would have estimated an $AaDO_2$ of approximately 45 +/-. Even if we move to the conservative side (45 - 15), an A-a gradient of 30 is substantial (and very close to the calculated $AaDO_2$ of 32.22torr).

D-dimer: the positive D-dimer in and of itself is not very impressive, but when coupled with an $AaDO_2$ suggestive of a shunt (greater than 20torr) and a history of acute onset trouble breathing, there is a very significant correlation to acute pulmonary embolism.

TEST TIP
Expected PaO_2 = FiO_2 X 5
$AaDO_2$ (A-a gradient) >20 suggests a shunt is present
positive D-dimer with an unexplained shunt suggests a pulmonary embolism

66. B ARDS frequently accompanies sepsis, especially in prolonged cases of such. ARDS should be suspected in cases of sepsis, pancreatitis, multi-system trauma, MODS and any other pathology that might precipitate hypoxia of the lung tissue itself.

67. B The patient's obesity and recent Roux-en-Y procedure should have made you highly suspicious of all options as this procedure requires significant bowel manipulation and incarcerated (trapped) gastrointestinal bleeding is very possible. The "pulmonary S2" should identify the pulmonary embolism as your first choice however.

TEST TIP
"pulmonary S2" ⇒ pulmonary embolism

68. A The AMI has probably damaged the left anterior wall and caused rupture of the papillary muscle affecting proper *closure* of the mitral valve. Because the muscle has ruptured and thus failed the mitral valve is allowed to prolapse resulting in significant pulmonary congestion. This is typically auscultated as a systolic murmur (the mitral should be closing during systole) at the apex of the heart (where the mitral valve is auscultated best, commonly referred to as the "PMI" or "Point of Maximal Impulse"). While congestive heart failure is a definite possibility, the systolic murmur is key to the question. Bezold-Jarish *reflex* is an actual physiologic phenomenon involving significant arrhythmias with significant drops in preload and/or perfusion of the AV node, while interesting, it is not pertinent to this scenario, furthermore the actual answer option is for a *syndrome* not a *reflex*. Aortic stenosis will cause a systolic murmur but it's PMI will be in the second intercostal space just right of the sternal border.

69. B The key to this question lies in the location of the infiltrates. Not the "right middle lobe" but the fact that the infiltrates are *only* in the right middle lobe, they're isolated. This suggests an infectious growth that originated there and has spread retrograde through the pulmonary tree becoming symptomatic only after the entire lobe has been involved.

TEST TIP
"consolidated infiltrates" ⇒ pneumonia

70. D While the body does respond to hypoxia with increased ventilatory effort this is actually somewhat slow and requires significant hypoxia in the normal individual before activating. Hypercarbia will result in increased ventilatory effort but in reality the elevated CO_2 level is not the most significant stimulus. The elevated CO_2 will combine with H_2O in the CSF and dissociate to hydrogen (H^+) and bicarbonate (HCO_3^-) via a carbonic anhydrase reaction. It is actually the H^+ that promotes the most significant and fastest changes in the brain's ventilatory centers. Recall our new paradigm, acid = H^+, not CO_2

71. D The *significant* slope noted in the plateau portion of the $EtCO_2$ waveform is demonstrating a delay in CO_2 delivery to the sensor itself, thus an exhalation obstruction is present. Of the choices offered, asthma and emphysema are most typical of this sloped waveform configuration. The normal $EtCO_2$ waveform should look like the diagram at right, with the C-D slope forming a very slight upward deflection.

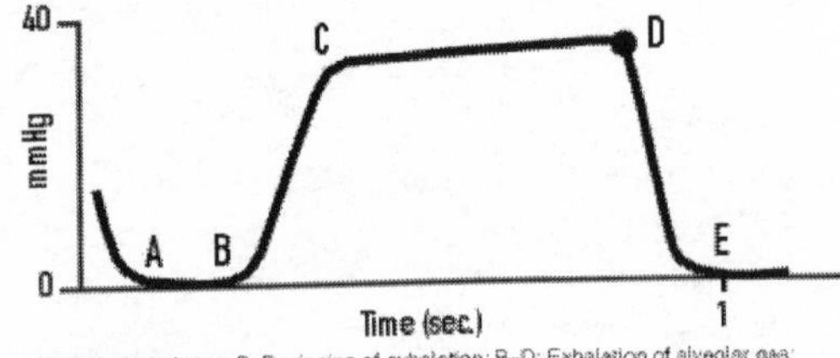

A: End of inhalation; B: Beginning of exhalation; B–D: Exhalation of alveolar gas; D: End exhalation and point of maximal or highest CO2 concentration {end-tidal CO2 (EtCO2)}; D–E: Inhalation.

72. C The waveform present is commonly referred to as a 'curare cleft' (referring to the use of curare related NMBA's in anesthesia at that time) and typically indicates the patient is awakening from their sedation/paralysis and attempting to breath before the ventilator was due to cycle the next breath. The most significant problem here is that the patient is making a sudden, significant respiratory effort through an endotracheal tube and ventilator circuit system which provides significant resistance unless the ventilator is in the appropriate mode with the sensitivity adjusted properly and/or pressure support available. Negative pressure pulmonary edema can ensue along with a myriad of other hemodynamic and intracranial pressure complications. Your first maneuver is to remove the patient from the vent minimizing resistance in the artificial dead space while quickly moving to re-sedate and/or paralyze. A and B are good choices but not the first course of action.

73. B See rationale for question 72 above.

74. B A partially obstructed exhalation valve will cause CO_2 trapping in the ventilator circuit, which may then be re-breathed. This results in an elevated $EtCO_2$ baseline. Other causes of elevated $EtCO_2$ baseline that may occur with typical transport ventilators include hyperpnea and tachypnea that exceeds the ventilator's support capability and an intake valve that is stuck open allowing exhaled CO_2 to enter the ventilator system only to be redistributed to the patient and thus 're-breathed'. Increased CO_2 production should raise the $EtCO_2$ value and plateau height but if the systems are working as designed, the baseline should not rise. Decreased spontaneous respiratory effort may also cause a rise in plateau level but if the ventilator has the appropriate settings and mode, even this may not occur, as the ventilator would supplement the decreased minute volume. Increased minute volume should decrease all $EtCO_2$ levels unless the ventilator system cannot handle the volume being moved by the patient; in that case the patient may re-breath exhaled gases assuming the circuit system would allow such. Transport vents do not utilize "CO_2 scrubbers" so discussion of CO_2 absorbent exhaustion is not pertinent. If you plan on using an anesthesia machine or going up in the Space Shuttle, see the appropriate reference materials.

75. B Infrared technology is only capable of measuring polar molecules with more than one element, i.e.- CO_2, N_20, etc. O_2 thus cannot be measure because it is a single element, non-polar molecule. Oxyhemoglobin however is a polar molecule and thus measurable with IR technology. Answer A, "oxygen saturation" is very vague and not as good a choice as B. Plasma O_2 partial pressure or PaO_2 must be measured using a co-oximetry system such as that found in an "ABG Machine". Carboxyhemoglobin levels traditionally have required a co-oximetry system to measure them however as of the writing of this text, new technology is being released which will allow for field carboxyhemoglobin level measurement using IR technology, if you answered 'D', give yourself a pat on the back for being 'on the cutting edge' of transport technology.

CARDIAC MANAGEMENT & HEMODYNAMIC MONITORING

1. Semilunar valves include;
   a. Tricuspid
   b. Aortic
   c. Mitral
   d. b & c

2. The patient's 12-lead ECG is demonstrating an inferoposterior wall AMI. The most likely site of the lesion is;
   a. proximal to mid LAD
   b. mid to distal circumflex
   c. distal PDA
   d. proximal to mid PDA

3. In ~90% of the population, the RCA is the primary blood supply to;
   a. left inferior wall
   b. septal wall
   c. right ventricle
   d. a & c

4. The first cardio-specific enzyme to elevate is;
   a. myoglobin
   b. CK
   c. CK-MB
   d. Troponin

5. The ECG finding that signals transmural ischemia is;
   a. ST depression
   b. ST elevation
   c. Q-wave development
   d. T-wave inversion

6. "Lateral leads" on the 12-Lead ECG are;
   a. I, II & III
   b. V1 & V2
   c. aVL & aVF
   d. none of the above

7. Your patient reports a prior history of CAD with chest pain reliably relieved by sublingual NTG. He has recently had recurring episodes of spontaneous chest pain in the morning upon waking. His condition may be described as;
   a. Prinzmetal's angina
   b. Stable angina
   c. Mixed angina
   d. Variant angina

8. Nitroglycerine's primary role in treating the AMI patient is;
   a. decreasing afterload
   b. decreasing sympathetic tone
   c. decreasing preload
   d. coronary vasodilation

9. Your patient takes verapamil daily. You should use caution administering which of the following;
   a. metoprolol
   b. diltiazem
   c. nifedipine
   d. all of the above

10. *Integrelin®* is;
   a. a low molecular weight heparin
   b. a fibrinolytic
   c. an plasminogenic inhibitor
   d. a glycoprotein IIb/IIIa inhibitor

11. An absolute contraindication to thrombolytics would be;
   a. HTN
   b. VT episodes
   c. CVA (11 months prior)
   d. AMI

12. Your patient is being transported with a PTCA sheath in the left femoral artery. Complications you should anticipate include;
   a. stagnant hypoxia of the right lower extremity
   b. pulmonary embolism
   c. LCA dissection
   d. femoral venous dissection

13. The patient's pacemaker is pacing constantly regardless of the patients intrinsic rhythm. This pacemaker failure would be described as;
    a. failure to pulse
    b. failure to synchronize
    c. failure to capture
    d. failure to sense

14. Your patient is exhibiting a diastolic murmur at the level of the mid-clavicular 5th intercostal on the left chest field. The most likely condition associated with this is;
    a. tricuspid stenosis
    b. mitral stenosis
    c. aortic regurgitation
    d. mitral regurgitation

15. Which of the following murmurs would be heard only during diastole?
    a. aortic stenosis
    b. mitral regurgitation
    c. pulmonic stenosis
    d. mitral stenosis

16. Which of the following murmurs would be heard only during systole?
    a. ventricular septal defect
    b. mitral stenosis
    c. tricuspid stenosis
    d. aortic regurgitation

17. Your patient is exhibiting a systolic murmur at the level of the second intercostal at the right border of the manubrium. The most likely condition associated with this is;
    a. ARDS
    b. pulmonic regurgitation
    c. IHSS
    d. CHF

18. Your patient is experiencing a systolic failure. A typical cause of this could be;
    a. hypertrophic cardiomyopathy
    b. dilated cardiomyopathy
    c. acute drop in afterload
    d. none of the above

19. Your patient has been diagnosed with a DeBakey Type II aortic aneurysm. Complications include;
    a. AMI
    b. CVA
    c. HTN
    d. all of the above

20. Your patient is experiencing a left ventricular diastolic failure. Treatment would include;
    a. afterload reduction with sodium nitroprusside
    b. preload increase with fluid bolus
    c. inotropic stimulation with dobutamine
    d. preload decrease with nitroglycerine infusion

21. Your patient is demonstrating 1mm of ST elevation in leads III & aVF. Which of the following therapies could prove hazardous;
    a. nitroglycerine administration
    b. glycoprotein IIb/IIIa administration
    c. heparin administration
    d. aspirin administration

22. Your patient is 6 months status post heart transplant. He has had difficulties with organ rejection throughout the course of his recovery. You're now transporting him for chest pain. He demonstrates the following vital signs; HR 42, BP 72/30, RR 30, SpO_2 85% and states he "can't breath". Which of the following therapies will not be helpful;
    a. dopamine infusion
    b. fluid bolus
    c. transcutaneous pacing
    d. atropine

23. The pressure bag on your arterial line set-up should be;
    a. inflated to 50mm Hg above the patient's systolic pressure
    b. observed to rise in pressure during descent for landing
    c. evacuated when performing fast flush tests
    d. monitored for drop in pressure while descending altitude

24. The transducer can;
    a. be placed at the phlebostatic axis for proper leveling
    b. be used to perform a fast flush test during pressure system checks
    c. opened to room air for zeroing of the system
    d. all of the above

25. An arterial line demonstrates a dicrotic notch. The dicrotic notch may represent;
    a. closure of the aortic valve
    b. closure of the pulmonic valve
    c. beginning of diastole
    d. all of the above

26. Slurring of the dicrotic notch would suggest;
    a. diastolic 'hold' phenomenon
    b. mitral valve disease
    c. aortic valve disease
    d. pulmonary hypertension

27. The correct chain of events in normal physiology are;
    a. renal pressure drops ⇒ angiotensin I formed ⇒ aldosterone released
    b. renal pressure drops ⇒ aldosterone release ⇒ angiotensin released
    c. renal pressure drops ⇒ angiotensin II formed ⇒ ACE utilized
    d. renal pressure increase ⇒ renin release ⇒ aldosterone released

28. ACE inhibitors interfere with;
    a. angiotensin II triggering aldosterone release
    b. angiotensin I conversion to angiotensin II
    c. angiotensinogen conversion to angiotensin I
    d. renin activation of angiotensinogen

29. Your 64 year old male patient presents complaining of chest pain and shortness of breath exhibiting the following vital signs; HR 98, BP 72/28, RR 28, SpO_2 92%, Temp 38°C. 12-lead ECG demonstrates acute anterolateral wall AMI. Therapy already in progress includes ASA administration, supplemental O_2 via NC at 6lpm, IV access with NS at TKO. Sublingual NTG was administered X 1 by first responders, nothing further. Select the best treatment from the choices below;
    a. O_2 at 10lpm via NRB, fluid bolus 500ml, initiate NTG drip at 10mcg/min
    b. Give 2nd NTG sublingual dose, establish 2nd IV NS at 250ml/hr
    c. Initiate dopamine infusion, if BP fails to respond, utilize norepinephrine or phenylephrine infusion titrating to a MAP of 60mmHg
    d. Initiate dopamine, then dobutamine infusions, consider intubation

30. Which of the following is not a 'normal' value?
    a. CVP 2-6mm Hg
    b. RVP 15-25mmHg
    c. PCWP 8-12mmHg
    d. all are "normal"

31. The PA pressure line waveform is exhibiting an obvious anacrotic notch with large sharp waveform deflections. The likely location of the catheter tip is;
    a. RV
    b. RA
    c. PA
    d. CVP

32. During transport it is your assessment that the PA catheter tip has moved to the RV. From the options below, you should;
    a. have the patient forcefully cough
    b. ask the patient to roll side to side or reposition themselves
    c. withdraw the catheter with the balloon down until a CVP waveform is seen
    d. all of the above

33. Match the following hemodynamic profiles with the possible etiologies listed in the two boxes below (A thru R).
There is commonly more than one possible etiology for each

Profile	Possible Causes
a. PAP 42/28	a.
b. CVP 0mmHg	b.
c. PCWP 15mmHg	c.
d. PCWP 4mmHg & PAP 40/22	d.
e. CVP 13mmHg	e.
f. PAP 12/4 & CVP 12mmHg	f.
g. CVP 30mmHg, PAS 32mmHg, PCWP 31mmHg	g.
h. CVP 0 mmHg, PAP 12/4, PCWP 2 mmHg	h.
i. CVP 0 mmHg, SVR 400, CI 5.0	i.
j. CVP 1 mmHg, SVR 1420, CI 2.3	j.
k. CVP 12 mmHg, SVR 1100, CI 1.4	k.
l. CVP 12 mmHg, SVR 1100, CI 1.4, PCWP 3 mmHg	l.

A. *Pulmonary hypertension*
B. *Mitral valve regurgitation*
C. *Tricuspid stenosis*
D. *Left ventricular AMI*
E. *Right ventricular AMI*
F. *Hypoxia*
G. *Aortic regurgitation*
H. *Mitral stenosis*
I. *Acute hemorrhage*
J. *ARDS*
K. *Sepsis*
L. *Anaphylaxis*
M. *Aortic stenosis*
N. *Fluid overload*
O. *Excessive vasodilator usage*
P. *Pericardial tamponade*
Q. *Neurogenic shock*
R. *Pulmonary embolism*

34. During transport the patient's PA waveform has changed morphology to a low amplitude "rolling" waveform. The most likely cause of this change is;
    a. inadvertent withdrawal of the catheter to the RV
    b. inadvertent migration of the catheter to a wedge position
    c. inadvertent withdrawal to a CVP position
    d. a normal event during positive pressure ventilation

35. PA catheters are typically designed for the distal balloon to be inflated with;
    a. no more than 1.5ml air
    b. no more than 0.15ml air
    c. only enough air to cause a waveform change on the monitor
    d. none of the above

36. Inadvertent 'wedging' of the PA catheter is a common problem;
    a. shortly after initial placement of a new catheter
    b. during patient repositioning and transport packaging
    c. a & b
    d. none of the above

37. Which of the following will not contribute to over damping of a transduced pressure line?
    a. descending to a lower altitude
    b. stopcock closure
    c. kinking of the pressure tubing
    d. air infiltration of the system between the patient and the transducer

38. Your patient has experienced an acute left anteroseptal AMI. Which of the following is not commonly associated with this pathology?
    a. left anterior fascicular hemiblock
    b. right bundle branch block
    c. "flash" pulmonary edema
    d. AV nodal ischemia

39. When assessing transduced pressure line readings, you should always take measurements;
    a. at end-exhalation with a spontaneously breathing patient
    b. at end-inhalation with a mechanically ventilated patient
    c. at end-exhalation with a mechanically ventilated patient
    d. a & c

40. When performing a fast flush test of a peripheral arterial line you note six obvious "bounces" of slowly decreasing amplitude after the flush. The arterial line's dynamics would be described as;
    a. 'whipping'
    b. 'over-damped'
    c. 'under-damped'
    d. 'cushioned'

41. A contraindication to the use of an IABP would be;
    a. abdominal aortic aneurysm
    b. peripheral vascular disease
    c. aortic valvular insufficiency
    d. all of the above

42. The tip of the intra-aortic balloon is typically;
    a. 2-4cm below the apex of the aortic arch
    b. at the apex of the aortic arch
    c. midway between the renal arteries and the left subclavian artery
    d. even with the phlebostatic axis

43. While caring for a patient with an IABP in place you note rust colored flakes in the IABP balloon supply tubing. This would indicate;
    a. Helium source contamination or leak
    b. IABP pressurization hydraulic leak
    c. IABP lubrication excess
    d. IABP balloon failure

44. The IABP works on the basis of;
    a. inflation during diastole with deflation during systole
    b. inflation during systole with deflation during diastole
    c. augmenting coronary supply while reducing ventricular workload
    d. a & c

45. The IABP may be used successfully;
    a. as a bridge device during lung transplants
    b. status post PTCA in patient's with multi-vessel disease
    c. as an assist device for ECMO therapy
    d. status post aortic valve repair

46. When timing the IABP, place the device in;
    a. 1:1 assist mode
    b. 1:2 assist mode
    c. 2:2 assist mode
    d. 2:1 assist mode

47. When transporting the IABP patient, it is not essential to carefully monitor which of the following;
    a. SpO_2 on the left hand
    b. SpO_2 on the lower extremity on the insertion side
    c. renal output
    d. CVP waveform

48. In the event of an IABP mechanical failure, you should;
    a. set the IABP to pressure mode
    b. set the ECG trigger to A-fib to compensate for erratic rhythm changes
    c. cycle the balloon once every 30 minutes manually
    d. cycle the balloon manually timing with the ECG waveform or A-line waveform when available

49. The IABP timing strip below demonstrates;

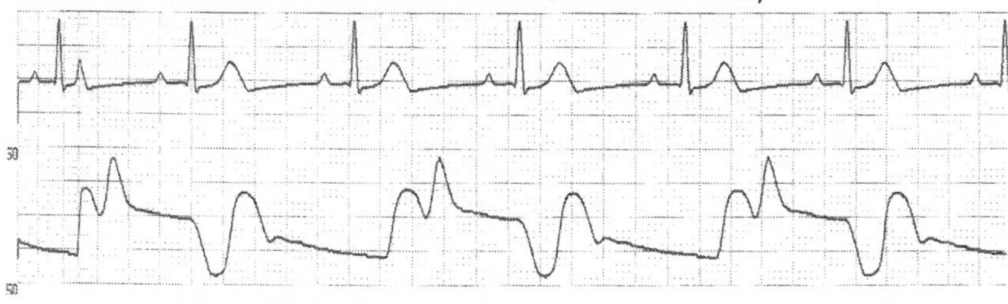

a. late inflation
b. early inflation
c. late deflation
d. early deflation

50. The IABP timing strip below demonstrates

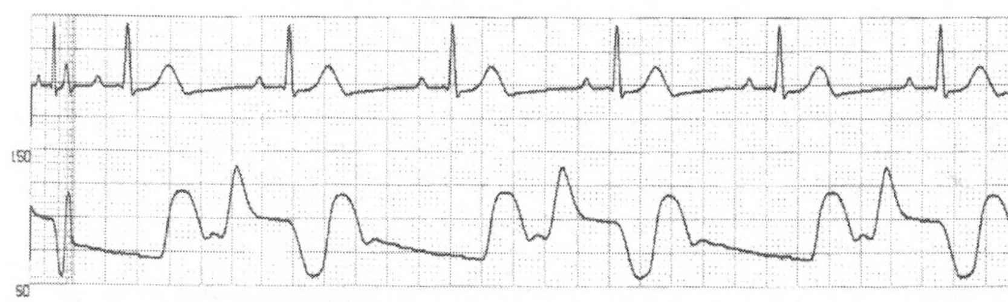

a. late inflation
b. early inflation
c. late deflation
d. early deflation

51. The IABP timing strip below should be corrected by;

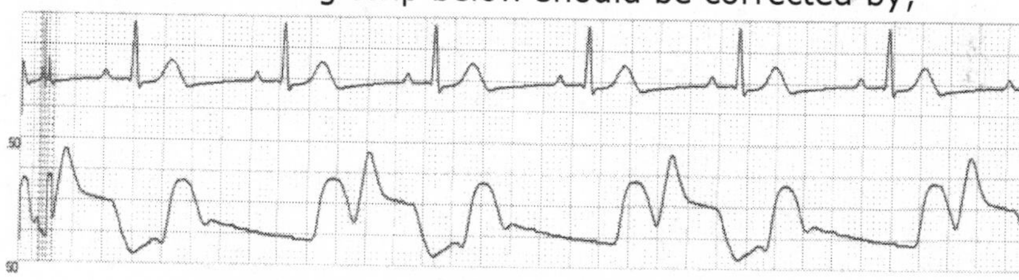

a. adjusting the inflation time earlier
b. adjusting the deflation time earlier
c. adjusting the inflation time later
d. adjusting the deflation time later

52. The ECG below is demonstrating;

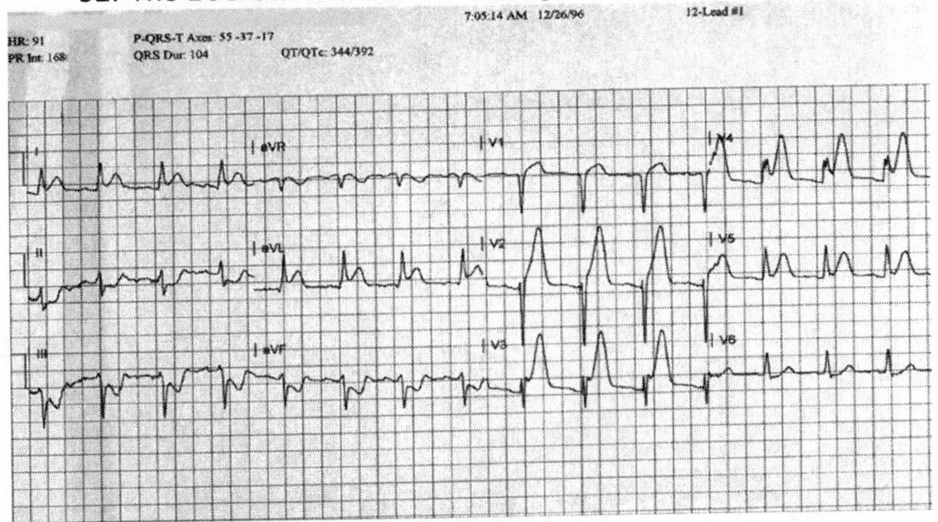

   a. inferoposterior wall AMI
   b. anteroseptal wall AMI
   c. lateral wall AMI
   d. apical AMI

53. The location of the infarct in the above question's patient is most likely located at;
   a. proximal RCA
   b. distal LAD
   c. proximal LAD
   d. distal circumflex

54. Complications associated with the AMI described by the previous two questions include;
   a. aortic stenosis
   b. left ventricular diastolic failure
   c. AV nodal infarction
   d. mitral papillary rupture

55. The ECG below is demonstrating;

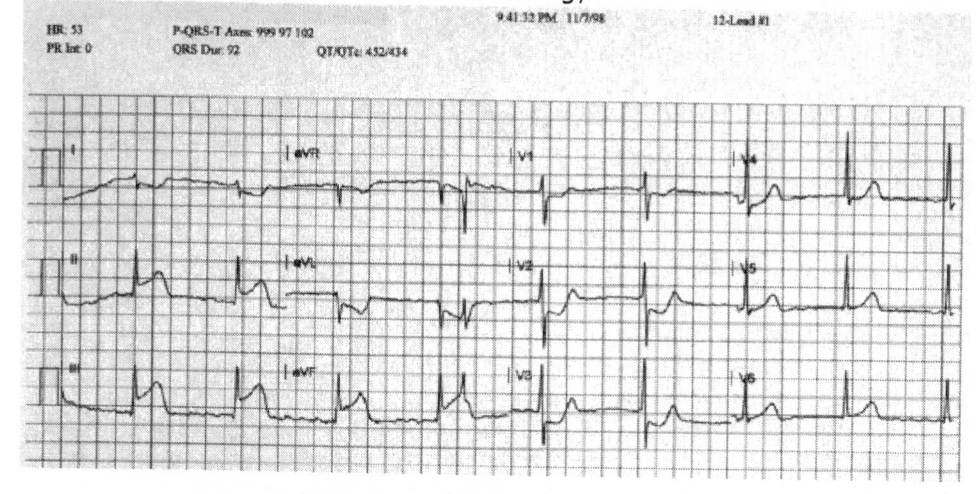

a. anteroseptal wall AMI
b. anterolateral wall AMI with septal extension
c. inferior wall AMI
d. inferoposterior wall AMI

56. Therapies with high potential to cause complications in the patient with the ECG in the prior question include;

a. sublingual NTG
b. maximizing FiO_2 to 1.0
c. isotonic fluid bolus of 250-500ml
d. aspirin 320mg PO

57. A 65 year old male patient is suffering from CHF. His CO is 3.9 and a peripheral arterial line is measuring a pressure of 112/67 with a heart rate of 94. He is mildly tachypneic on 6lpm of supplemental O_2. Further examination reveals a RAP of 9mmHg. An increase in which of the following would be expected with this patient?

a. plasma colloid osmotic pressure
b. interstitial colloid osmotic pressure
c. arterial hydrostatic pressure
d. vena caval hydrostatic pressure

58. A 79 year old female with end-stage emphysema has been diagnosed with cor pulmonale. She is able to maintain a saturation of 84% on 4lpm of supplemental O_2. Which of the following would not be an anticipated finding with this patient?
    a. atrial fibrillation
    b. low 2-3DPG levels with a normal H&H
    c. elevated RAP
    d. cardiomegaly on AP chest radiograph

59. In which type of shock would you anticipate an increase in cardiac output?
    a. septic shock
    b. hemorrhagic shock
    c. anaphylactic shock
    d. neurogenic shock

60. The lab finding most indicative of true shock is;
    a. low PaO_2
    b. high $PaCO_2$
    c. elevated K^+
    d. elevated lactate

61. Coronary blood flow is dependent upon;
    a. diastolic blood pressure in the aorta
    b. left ventricular end-diastolic pressure
    c. heart rate
    d. all of the above

62. The coronary arteries are perfused during;
    a. systole
    b. diastole
    c. both systole and diastole
    d. unable to answer

63. Cardiac output is determined by which of the following;
    a. CO= SVR X HR
    b. CO= AoDP – LVEDP
    c. CO= SBP – 2/3(DBP)
    d. CO= SV X HR

64. Stroke volume is determined by;
    a. heart rate, SVR and EF
    b. preload, afterload and contractility
    c. preload, SVR and heart rate
    d. preload, SVR and wall tension

65. Factors influencing left ventricular oxygen supply include coronary artery anatomy, diastolic pressure, oxygen extraction and;
    a. heart rate
    b. PA systolic pressure
    c. SVR
    d. oxygen demand

66. The primary determinant of myocardial oxygen consumption is;
    a. heart rate
    b. blood pressure
    c. myocardial wall tension
    d. contractility

67. The best indication for intra-aortic balloon counter-pulsation therapy is;
    a. circulatory support in post CABG patients
    b. acute AMI with cardiogenic shock
    c. all patients status post AMI
    d. patients status post successful PTCA

68. The most common complication associated with intra-aortic balloon counter-pulsation therapy is;
    a. ischemia of the limb distal to the insertion site
    b. aortic dissection
    c. thrombocytopenia
    d. infection/septicemia

69. Which of the following IABP timing errors is potentially most harmful?
    a. early inflation
    b. early deflation
    c. late inflation
    d. late deflation

70. The primary trigger for the IABP is typically the;
    a. dicrotic notch
    b. arterial pressure catheter
    c. ECG
    d. helium pressure

71. Which law will prompt the IABP to purge on ascent while transporting by air?
    a. Charles's Law
    b. Boyle's Law
    c. Graham's Law
    d. Gay-Lussac's Law

72. When the balloon pump is purging, the transport provider needs to anticipate which of the following?
    a. bradycardia
    b. bradypnea
    c. increased urinary outflow
    d. hypotension

73. You are transporting a patient with an IABP in place functioning in a 1:1 assist mode. You note the pump making irregular sounds as opposed to the previous rhythmic sounds. Your patient has entered a rather erratic atrial fibrillation and the pump appears to be having difficulty timing. Which of the following would be most likely to assist the device with optimal timing coordination?
    a. turn down the ECG gain
    b. change the assist ratio to 1:2
    c. lower the balloon volume to 75% it's initial volume
    d. set the IABP to an internal or pressure trigger

74. Which of the following statements is true regarding the anatomy of the femoral vein?
    a. The femoral vein is medial to the femoral nerve and lateral to the femoral artery.
    b. The femoral vein is medial to the femoral artery and lateral to the femoral nerve.
    c. The femoral vein is lateral to the femoral artery and nerve.
    d. The femoral vein is medial to the femoral artery and nerve.

75. When cannulating the subclavian vein, the most common "significant" complication is;
    a. ventricular arrhythmias
    b. pneumothorax
    c. catheter misdirection to the internal jugular
    d. failed access

76. You are about to transport a 53 year old male with a self inflicted gun shot wound to the left chest. Fragments of the bullet have resulted in an aortic aneurysm that is 'leaking' and a left pneumothorax. The patient has already received a chest tube to the left chest that appears to be functioning properly. The patient's vitals are BP 121/77, HR 104, RR 24, SpO_2 99% on non-rebreather. You will be transporting from Tulsa, Oklahoma to Houston, Texas. Your team and the sending facility agree the patient would benefit from an arterial catheter and central venous catheter for monitoring during flight. Your best site for central venous access would be;
    a. right internal jugular (IJ) vein
    b. left subclavian vein
    c. femoral vein
    d. right subclavian vein

77. You are transporting a patient from a "diagnostic cath lab" to a facility with an interventional team and cardiothoracic surgical backup. During the "diagnostic procedure" they recognized multi-vessel disease and encountered "difficulties" according to the records. They inserted a Swan-Ganz catheter and supposedly wedged the catheter at a depth of 85cm. The Swan is inserted via the right IJ through a cordis. The typical depth where this PA catheter should have wedged is around;
    a. 30cm
    b. 40cm
    c. 50cm
    d. 60cm

78. You are initiating transport from another "diagnostic cath lab", again to a facility with an interventional team with surgical backup. You note that the patient has consistent bleeding from IV sites, the foley and a cut lip (from a traumatic intubation in the ED). You should first suspect;
    a. DIC
    b. dilutional thrombocytopenia
    c. prior aspirin use/abuse
    d. heparin toxicity

79. Based on the situation above in question 78, a review of the records indicates you were right, the patient was not properly "reversed" and you receive orders to administer protamine sulfate. Complications of this drug's administration include all but which of the following;
    a. bradycardia
    b. hypotension
    c. bronchoconstriction
    d. pulmonary hypertension

80. A normal ejection fraction (EF) is;
    a. 50-55%
    b. 60-65%
    c. 75-80%
    d. > 90%

81. The ECG below demonstrates;

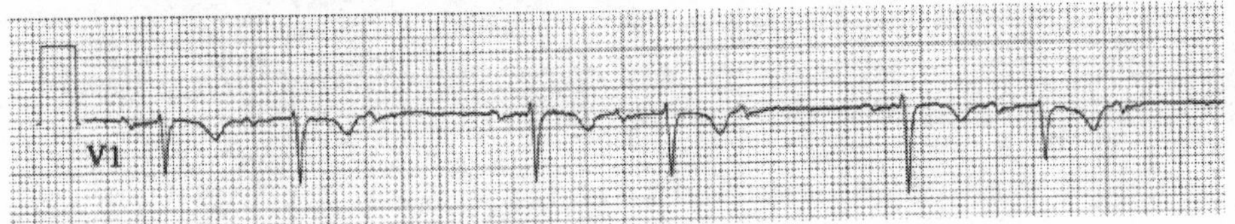

    a. normal sinus rhythm
    b. 1st degree AV block
    c. 2nd degree AV block type I
    d. 2nd degree AV block type II

82. The ECG below demonstrates;

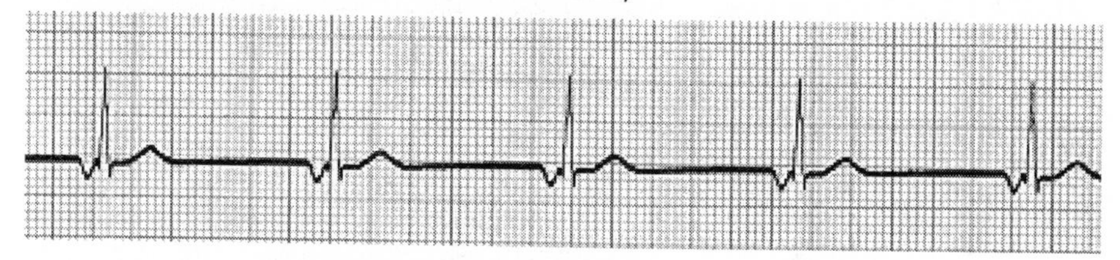

a. normal sinus rhythm
b. PJC's
c. junctional escape rhythm
d. accelerated junctional rhythm

83. Which of the following best describes the ECG below?

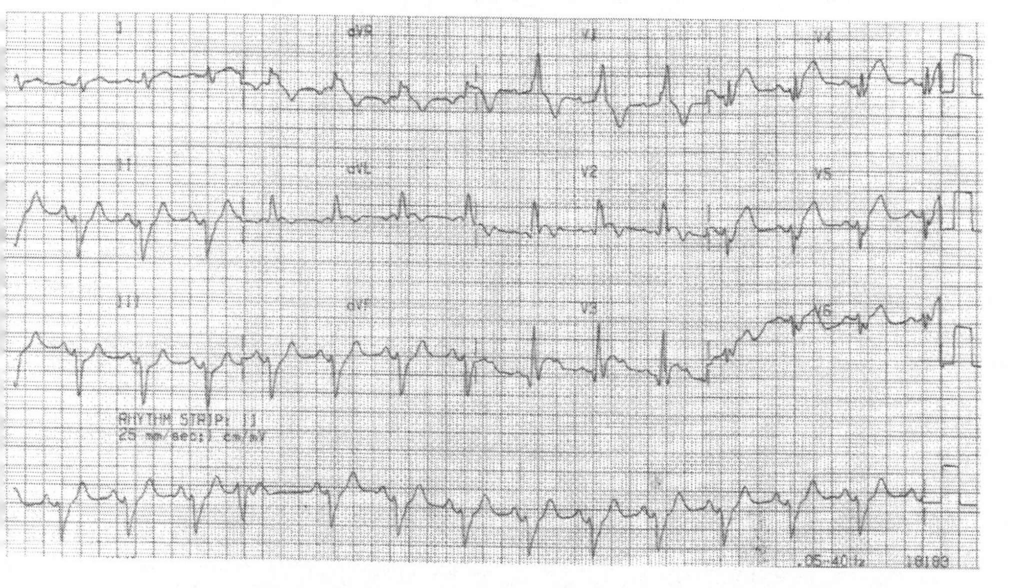

a. normal sinus rhythm with 1st degree AV block
b. sinus rhythm with bifascicular hemiblock (RBBB and LAFHB)
c. accelerated idioventricular rhythm with RBBB
d. sinus tachycardia with LBBB

84. You are transporting a post cardiac cath patient to a tertiary care facility. The patient has an arterial line, central line and PA catheter in place. Vitals are currently within normal limits and appear to be stable. While transporting your patient to the airfield from the sending facility you notice the following ECG pattern begin to occur. Which of the following treatments would be most appropriate?

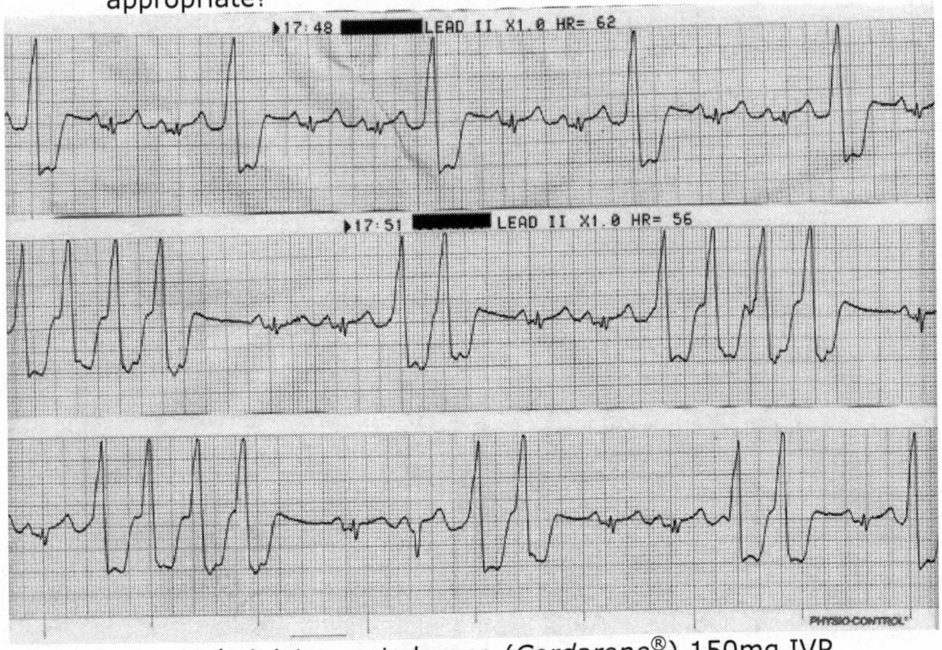

a. administer amiodarone (*Cordarone®*) 150mg IVP
b. verify the arterial waveform has not changed and check the level of the arterial transducer
c. verify the depth of the PA catheter, if changed, consider pulling the PA to a CVP position
d. begin the necessary steps to intubate the patient in preparation for further arrhythmias and deterioration

85. The following ECG demonstrates;

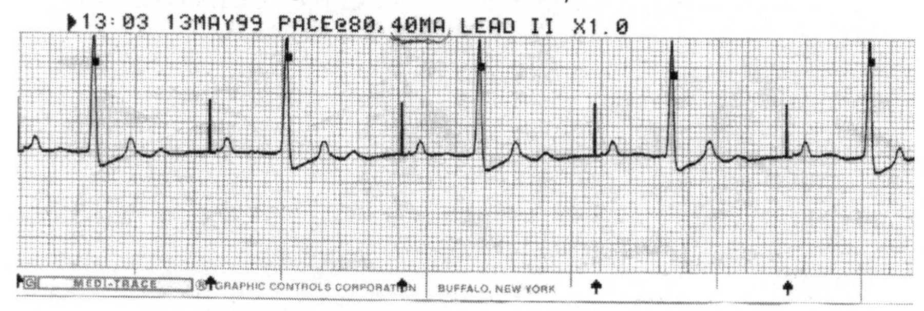

a. bigeminal PVC's
b. AV dissociation
c. paced rhythm with latent capture
d. paced rhythm with failure to capture

86. Your patient's ECG demonstrates the rhythm below during transport. Which medication is most likely to correct the arrhythmia per current standards of care?

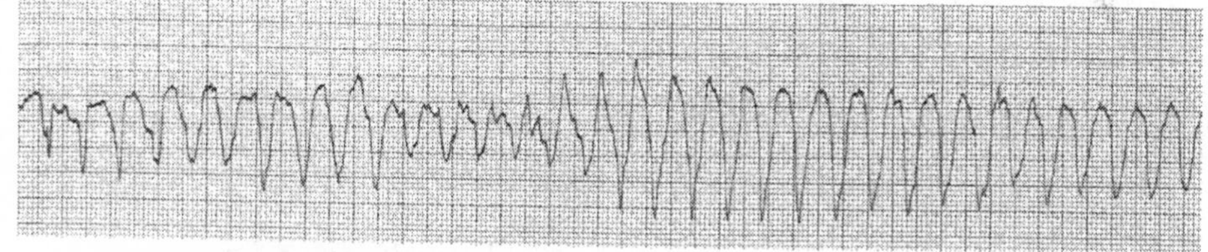

a. lidocaine
b. amiodarone
c. procainamide
d. magnesium

87. During transport you note that your PA catheter's distal port suddenly begins demonstrating the following tracing and the insertion depth at the introducer/cordis hub has not changed. Your first strategy for correcting this would be;

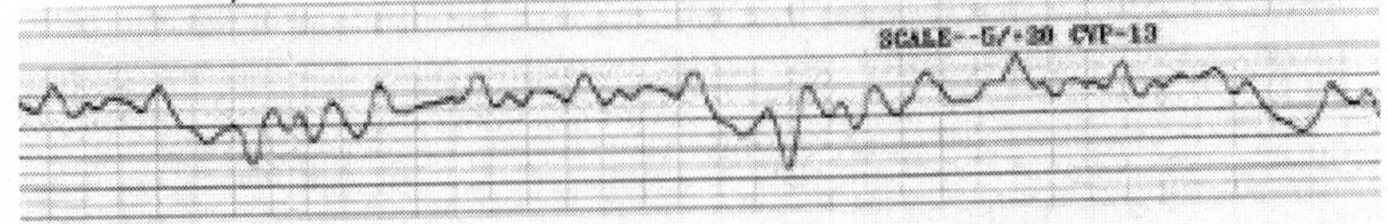

a. verify the balloon is up and then re-deflate
b. verify the balloon is down then ask the patient to cough forcefully
c. verify the balloon is down then advance the catheter until a PA waveform is seen
d. verify the balloon is up and advance the catheter until a PA waveform is seen

88. While observing the sending physician place a Swan-Ganz catheter, you observe the tracing below. What has just happened as evidenced by the PA tracing?

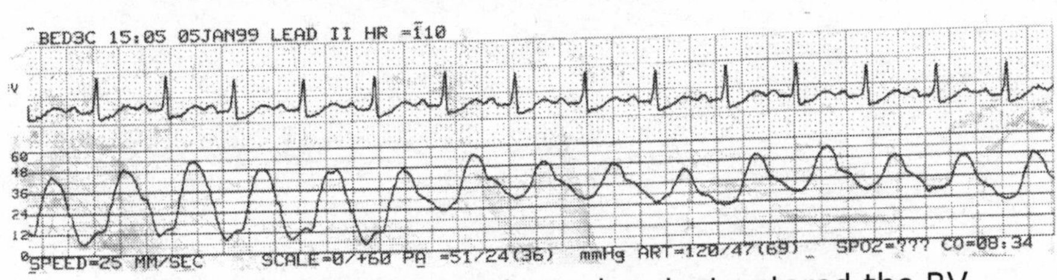

a. The tip of the catheter has just entered the RV
b. The tip of the catheter has just entered the PA
c. The tip of the catheter is wedged
d. The tip of the catheter has just coiled and re-entered the RA

89. Reference the strip in the prior question, there is an obvious anacrotic notch on the first six waves of the Swan's tracing. The ascending waveform just prior to the anacrotic notch would represent;
a. atrial filling
b. atrial kick
c. mitral valve opening
d. aortic stenosis

90. Reference the strip in question 88, the last 9 waves of the Swan tracing seems to have a slightly wandering or varied baseline. This is commonly caused by;
    a. natural variation due to breathing
    b. natural resonance of the transducer system
    c. equipment inaccuracy and expected until the system is re-zeroed
    d. the paper feeding into the printer irregularly

KEY & RATIONALE

CARDIAC MANAGEMENT & HEMODYNAMIC MONITORING

1. B The tricuspid and mitral valves lie between atria and ventricles, thus they are atrioventricular valves. The semi-lunar include the pulmonic and aortic valves.

2. D The inferior and posterior walls of the left ventricle are supplied primarily by the right coronary artery in the vast majority of the population (~90%). Infarction in both areas suggests a higher lesion or occlusion site. A proximal to mid LAD lesion would typically cause an anteroseptal wall AMI. Occlusion of the circumflex is typically associated with lateral wall AMI.

3. D See previous rationale. The RCA supplies (in order of flow from the aorta); the SA node (55% of the population), the RV, the AV node (90% of the population), left posterior wall and left inferior wall (90% of the population).

TEST TIP

RCA becomes the PDA and supplies the SA node, right ventricle, AV node, left ventricular posterior and left ventricular inferior walls.

LCA divides primarily into the LAD and Circumflex. The LAD feeds the left ventricular anterior and septal walls including the right bundle branch and left anterior fascicle. The circumflex feeds the left ventricular lateral wall.

The left posterior fascicle typically has three blood supplies, two from the LAD and one from the RCA.

This is the basic anatomy for the majority of the population, obviously exceptions do exist.

4. C While myoglobin will elevate first, it is not "cardio-specific". Total CK will also elevate with CK-MB but again may be caused my non cardio-specific causes (i.e.- CK-MM and CK-BB) Troponin, while very cardio-specific is slower to rise than CK-MB.

5. B ST elevation is the hallmark of transmural ischemia. ST depression may signal subendocardial injury or simply be a reciprocal finding of ST elevation elsewhere. Q-waves are definitive of infarcted tissue and present late in the AMI process. T-wave inversion is a very early sign indicative of early injury.

6. D The lateral leads are I, aVL, V5 & V6. While answers A. and C. include parts, they also include incorrect leads. The "best" answer would be D.

TEST TIP

Inferior Leads	**II, III, aVF**
Septal Leads	**V1, V2**
Anterior Leads	**V3, V4**
Lateral Leads	**I, aVL, V5, V6**
Posterior Leads	**V1-4(reciprocal findings), V7, V8, V9**
Right Ventricular Leads	**V1R-V6R**

See diagram p.117

7. C This patient is describing a history of stable angina with recent onset of variant angina. When these occur simultaneously it is referred to as "mixed angina".

8. C Nitroglycerine is a pre-load reducing agent and it is via this mechanism that cardiac stress is relieved. Coronary vasodilation was one of the first theorized mechanisms but has since been proven unsubstantiated. NTG plays no direct role in sympathetic tone but may reduce it secondarily as chest pain is relieved from the preload reduction.

TEST TIP
"Nitroglycerine is a preload reducing agent"

9. D Verapamil (*Calan®*) is a cardioselective calcium channel blocker. As such it promotes coronary vasodilation and reductions in dromotropy and inotropicity. Adding additional calcium channel blockers (diltiazem & nifedipine) or beta-blockers (metoprolol) presents a hazardous situation in that calcium channel blocker overdose may occur or in the case of the beta-blocker, complete blockade of primary mechanisms involved in stimulating the heart.

10. D Eptifibatide (*Integrelin®*) is a Glycoprotein IIb/IIIa inhibitor used to disrupt the clotting cascade. Other medications in this category include abciximab (*Reo-Pro®*) and tirofiban (*Aggrastat®*)

11. C Any type of CVA within the last 12 months is an absolute contraindication to the use of thrombolytics for AMI. HTN is a relative contraindication and dysrrhythmias do not present any contraindication per se.

12. C Complications associated with femoral sheaths post PTCA include stagnant hypoxia of the leg where the sheath is placed, bleeding and sheath occlusion. Other complications such as coronary artery dissection, pericardial tamponade, dysrrhythmias and AMI must be anticipated from the PTCA that was performed via the sheath.

13. D The pacemaker malfunction described is asynchronous pacing. It is not sensing the patient's intrinsic rhythm to synchronize with. Asynchronous *mode* would cause the same problem but that was not offered as an answer option. Failures to pulse and synchronize are distracters and placed early in the choice selection for added distraction. Failure to capture would indicate the pacer is firing but not prompting cardiac depolarization and/or mechanical contraction.

14. B The 5th intercostal space, mid-clavicular line on the left side, is the point of auscultation for the mitral valve, also called the point of maximal impulse or PMI. Hearing a murmur here identifies the mitral valve as the likely source. Because the murmur is diastolic one must consider what the mitral valve is doing during diastole (it should be opening). A stenotic mitral valve will cause the murmur during the opening phase. Mitral regurgitation would be heard at the same location but it would occur during systole.

15. D See explanation above. As you work through the answer options you must ask yourself with each disease, "What is this valve doing during diastole?" Stenosis is associated with opening, regurgitation or incompetence is associated with closing of the mitral valve.

16. A Ventricular septal defect would be heard during systole as blood preferentially shunts from the higher pressure ventricle to the lower pressure ventricle (left to right shunt). If VSD was unfamiliar to you, the process of elimination should rule out the other options.

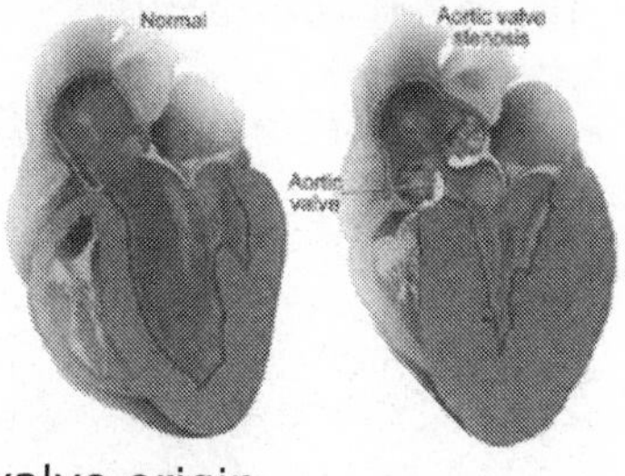

17. C IHSS or Idiopathic Hypertrophic Subaortic Stenosis is associated with hypertrophic cardiomyopathy and aortic valve stenosis (see picture right). ARDS and CHF may precipitate other murmurs over a long period. The site of auscultation should direct you to an aortic valve origin.

18. B A systolic failure signifies a failure to 'clear' or adequately evacuate the ventricles of blood during systole. Patients with *hypertrophic* cardiomyopathy typically have normal to elevated systolic clearance; instead they are unable to load, or fill the ventricle with blood adequately during diastole. Patients with a *dilated* cardiomyopathy are typical of CHF or status post AMI victims. The ventricle, unable to clear during systole, experiences volume overloading, which stretches or 'dilates' the ventricle. An acute drop in afterload would not prompt a systolic failure; in fact dropping afterload is commonly a treatment for *dilated* cardiomyopathy.

19. D A DeBakey Type II aortic aneurysm occurs in the ascending arch of the aorta. As such the dissecting aortic lumen may progress upstream invading the aortic ostia and coronary inlets causing an AMI. If the dissecting lumen invades the arch further downstream, occlusion of the carotid artery may occur prompting CVA. Because of the pain and hemodynamic changes caused by the aneurysm, HTN may be resultant. Primary HTN may contribute to precipitating the aneurysm initially.

20. B Diastolic failure indicates the ventricles are failing to adequately fill prior to systole. Afterload reduction will assist clearing of the ventricles, not filling. Inotropic stimulation is not indicated to assist filling, typically failed clearing or systolic failure would indicate inotropes. Decreasing preload further with NTG will only antagonize the scenario. Beta-blockade might be indicated to slow the heart rate and prolong the diastolic filling time but that was not offered as an answer choice.

21. A ST elevation in III and aVF suggest inferior wall infarction. The inferior wall is supplied primarily by the RCA in the majority of the population. Because the RCA also supplies the right ventricle prior to feeding the left inferior wall, it should be assumed that the RCA lesion is high and causing a right ventricular infarct (RVI) until ruled out. Assuming there is an RVI you must also assume pre-load is already compromised and administration of nitroglycerine can exacerbate the situation. GIIb/IIIa inhibitors and heparin are commonly given to AMI victims and shouldn't present an unusual risk with an RVI patient. Aspirin can be a hazardous medication in bleeding disorders and some other specific populations but the question doesn't offer any of those qualifiers.

TEST TIP
Inferior wall AMI frequently is accompanied by left posterior wall and right ventricular infarct (RVI). Use caution with nitrate administration.

22. D This question is typical of the 'distraction' variety. "Heart transplant" is really all that's needed to identify the question's answer and all other data is simply a distraction here. Transplanted hearts cannot have the vagus nerve (CN10) reattached; hence atropine will have no effect on the heart rate of this patient.

TEST TIP
"Transplanted hearts don't respond to atropine"

23. D The pressure bag should always be pressurized to 300mmHg regardless of patient BP. The transducer systems are designed for this constant pressure. During descent, atmospheric pressure on the bag would rise therefore compressing the volume in the bag and dropping the bag pressure. When zeroing the system, the transducer or tubing must be opened to the atmosphere but the pressure bag isn't altered. The stopcock between the transducer and the patient is turned off to the patient and opened to the atmosphere while zeroing.

24. D The transducer must be placed at the level of the RA or the phlebostatic axis (4th intercostal, mid-axillary line) during hemodynamic monitoring for accurate readings. The transducer is typically the location of the fast flush 'pig-tail' (see right) or clamp and also typically has a stopcock, which allows for zeroing of the monitor.

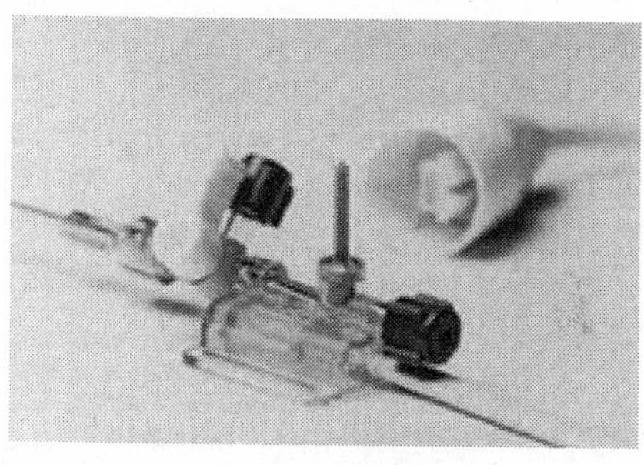

TEST TIP
Hemodynamic transducers should be placed level with the right atria at the "phlebostatic axis" (4th intercostal, mid-axillary line)

25. D The dicrotic notch represents closure of the last heart valve upstream. Thus a peripheral arterial line's dicrotic notch represents closure of the aortic valve vs. a pulmonary artery catheter's dicrotic notch will represent closure of the pulmonic valve. The pulmonic and aortic valves only close as a result of reduced pressure upstream in the ventricle. For purposes of these exams and this discussion, diastole begins at the dicrotic notch. *In fact diastole begins slightly before the dicrotic notch while the heart's pressures and volumes move through a brief phase called the "isovolumetric relaxation" in which pressure changes but volume does not. For further readings you would want to research "Wigger's Diagram", (this is not addressed directly or otherwise on the CFRN, CEN or FP-C exams as of the writing of this book)*

26. C Slurring of the dicrotic notch suggest that the valve represented by that particular arterial waveform has an insufficiency allowing regurgitation to occur. The actual notch with it's immediate slight rise or 'bounce' in pressure is caused by the valve closing as fluid attempts to backflow into the ventricle from which it was just ejected. The valves are pliable enough that they typically close, bulging into the ventricle as fluid attempts to backflow past them, then rebound due to their elasticity causing a slight, momentary forward surge in the fluid of the artery. This momentary, slight surge causes the notch pattern seen on the waveform. Diastolic 'hold' phenomenon is an imaginary term used as a distracter. Mitral valve disease and pulmonary hypertension while directly involved with PA catheter pressures do not play a role in slurring of the dicrotic notch.

27. A The RAA cascade or Renin-Angiotensin-Aldosterone system is critical to fluid and pressure management throughout the body. The system is triggered typically by a reduction in renal perfusion and glomerular filtration rate (GFR). As the GFR is reduced the flow through the nephrons of the kidney is reduced allowing more time for sodium reabsorption in the thick ascending loop. The macula densa at the juxta glomerular apparatus (JGA) senses the lack of sodium in the filtrate at the top of the thick ascending loop, which will eventually become urine. In response to the low sodium, the cells of the JGA release renin into the bloodstream leaving the kidneys. Renin acts as a catalyst to trigger angiotensinogen in the blood stream to convert to angiotensin I. Angiotensin I will convert to angiotensin II in the presence of angiotensin converting enzyme (ACE). The majority of ACE is located in pulmonary epithelial tissue. Angiotensin II upon leaving the lungs will immediately promote significant vasoconstriction and upon returning to the kidneys prompt the release of aldosterone. Aldosterone will function to trigger mechanisms, which increase the activity and expression of sodium-potassium pumps in principle cells of the distal convoluted tubules and collecting ducts of the nephrons. It will also trigger the creation of additional sodium reabsorption channels in these same cells. All of these mechanisms serve toward the common goal of increasing blood pressure to the kidneys in hopes of increasing the glomerular filtration rate.

28. B See explanation of RAA system above. When attempting to recall ACE inhibitors think of the 'prils'. benazepril (*Lotensin*®), captopril (*Capoten*®), enalapril (*Vasotec*®), lisinopril (*Prinivil*®), etc...

29. D This question is complex in that a large amount of data is presented which should be factorial when determining the answer. Furthermore the answer choices presented are somewhat incomplete and/or vague. The strategy of choice here may be to identify what therapies you don't want to provide, as they will obviously cause further harm to the patient. One such therapy would be decreasing the BP of the patient at this time. This patient is presenting with an AMI and secondary cardiogenic shock. Of the choices offered, A & B include nitroglycerine, a possible choice later on in this patient's therapy but not ideal at this time. Option C begins with a good plan but deteriorates quickly when failure to respond to the dopamine is to administer norepinephrine or phenylephrine. Both of these later agents are α-1 agents that will increase afterload substantially. While pressure is a concern here, the causation is a failing pump, not vasodilation. With A, B & C ruled out, option D, while aggressive and rather vague is the only option that doesn't suggest an obvious harmful element to the patient at hand.

30. B The RVP is an arterial waveform and thus must have a systolic and diastolic reading vs. the CVP and PCWP, which are mean pressures, are represented by a single number. Answer B states that a single number range of 15 to 25 is normal. This would be accurate if the choice read that the RV systolic pressure range is 15-25mmHg. The CVP and PCWP readings are the norms utilized on the exams as of this writing. D is offered as an attractive distracter, a "sucker punch" if you will. This question placed at the end of the exam, when you are mentally fatigued would prove hazardous for most test takers. This emphasizes the need to practice taking exams for endurance. Stay sharp!

31. A An *anacrotic* notch is a notch on the ascending side of the waveform deflection (vs. a dicrotic notch, which is found on the descending side of the waveform) (Both notch types are seen on the waveform at right). The anacrotic notch is identified as a notch to the left of the waveform's peak. When a PA catheter is placed it must pass through the RV (see lower right) at which time an anacrotic notch should be noticed which represents "atrial kick". Atrial kick provides us with ~22%-33% (depending on text referenced) of our cardiac output, hence the urgency in correcting atrial fibrillation in patients that are status post heart bypass ("a fresh heart").

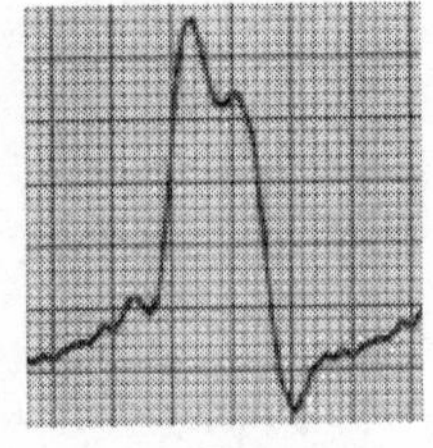

32. C Again, this question offers some "attractive distracters". Answers A and B may sound like a good choice, and they are.... for an inadvertently wedged PA catheter. This PA catheter isn't wedged; it's been pulled/ withdrawn to a more shallow position in the RV. While the catheter tip is in the RV it will likely irritate the ventricle and may precipitate ventricular arrhythmias. Two options exist, advance the catheter back into the PA or withdraw the catheter to the RA/CVP position. Advancing of the PA catheter is typically not considered a "nursing function" and left to the advanced practitioner or physician. This is a moot point here since the option of advancement was not offered in the answer selections. Be sure the balloon is always *down* <u>prior to</u> pulling back.

33.a. A, B, D, F, G, H, J, M, N, R All of these options will promote an elevated PA pressure. The ideal way to approach this question is to simply run down the list of etiologies offered and ask yourself, "With this condition, would it cause fluid to back up from it's disease location to the pulmonary artery and reflect a higher pressure?" or, "Will it cause vasoconstriction in the pulmonary artery while maintaining a normal fluid volume?" A. is a "gimme". B tells you that the high pressure left ventricle is pushing fluid backwards through the mitral valve due to it's regurgitation/ incompetence, as such high pressures would occur in the pulmonary system. D would cause failure of the left ventricle to clear properly. Meanwhile the right ventricle would continue pumping the same volume into the pulmonary system. The backup would cause a fluid overload in the lungs. F isn't quite as intuitive. Remember the lungs respond to hypoxia by vasoconstriction

(HPVR) to shunt the blood to the areas of the lung that are not hypoxic. If the entire lung is hypoxic, then the entire lung vasoconstricts resulting in an elevated PA pressure. G would cause an elevated PAP via the same mechanism as option B just a little further downstream. H, mitral stenosis tells us the mitral valve doesn't want to open. The right ventricle never gets the memo so it keeps pumping the same volume into the pulmonary tree, meanwhile the mitral valve isn't letting it escape from the left atria into the left ventricle and pressure builds. J, ARDS will cause regional areas of the lung to have profound hypoxia and fluid shifting. As such pulmonary pressures skyrocket. Interestingly, ARDS is frequently accompanied by sepsis so it may be surprising that the CVP may be low as well as the CI and SVR. This is one specific case when a PA catheter may really confirm the presence or lack of a disease state. M, Aortic stenosis will work via the same mechanism as H, again just further downstream. N, again is a "gimme". R, or pulmonary embolism again is not entirely intuitive. A PE downstream of the catheter may create a substantially elevated PA pressure. If the PE is located in another branch of the pulmonary artery or not large, the subsequent hypoxic vasoconstriction and shunting effect may be minimized enough to cause a normal or minimal elevation in PA pressure.

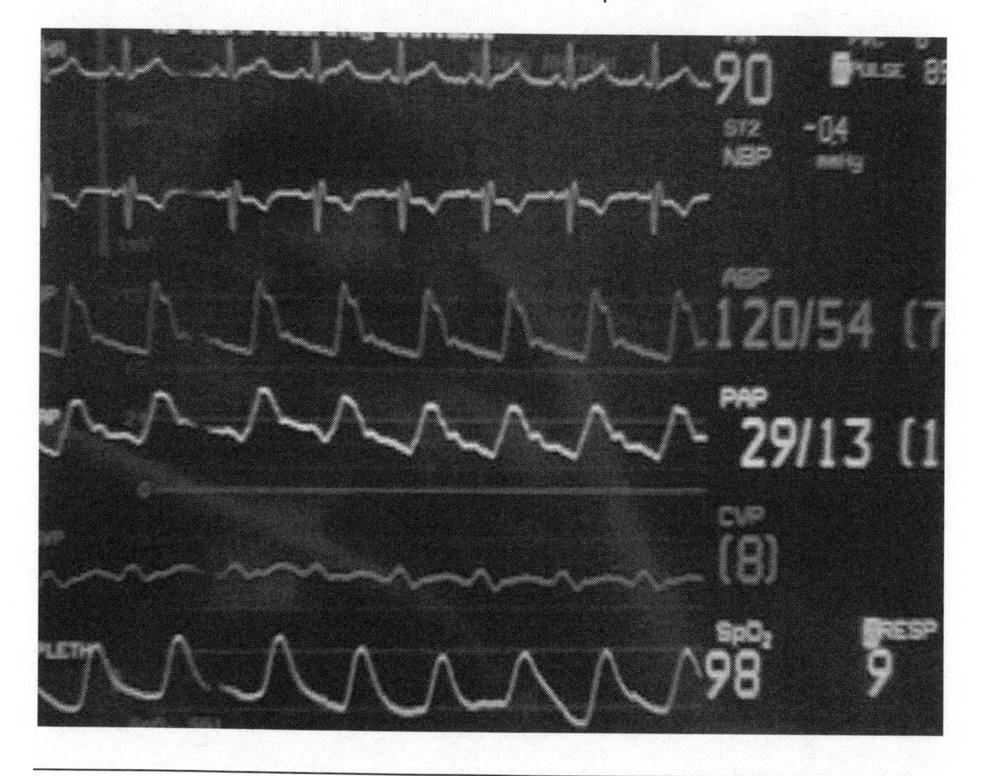

Consider using the diagram below to form a mental picture of the cardiovascular system that is easy to troubleshoot. This is my 'heart made easy' with a PA catheter placed and wedged. Imagine various components of the system failing to open, close, contract or relax and then decide what the pressures would be like in the other areas. I.e.- if we imagine the right ventricle is failing to contract due to a right ventricular AMI, we would anticipate the right ventricle (RV) would fail to 'clear' and pressures inside that ventricle would climb. Pressure would 'back up' upstream, like damming a river. Right atrial pressures would climb, CVP would climb. Downstream of our 'dam' we would see lower pressures. Lower PAS, PAD, PCWP, etc. etc.

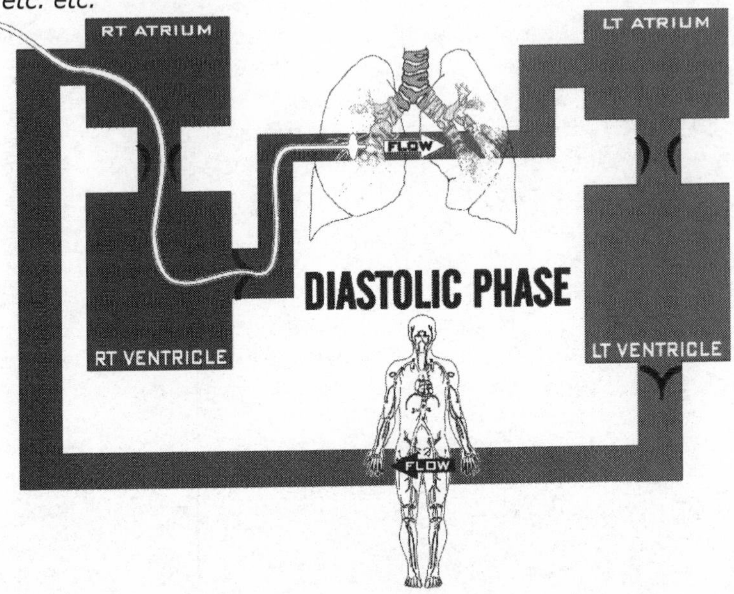

33.b. I, K, L, O, Q The CVP is low. Look for anything that would cause excessive fluid loss or pooling in the peripheral vascular regions such as in a redistributive shock state. ARDS may look like a good choice and it is typically accompanied by sepsis however, ARDS alone would not cause a low CVP typically.

33.c. B, D, G, H, M, N, P You should take note that many of the same choices were made for the elevated PAP (a.). We added only one more, P. pericardial tamponade. So why was pericardial tamponade not a choice with a PAP of 42/28? The hallmark of a tamponade is that the CVP, RV, PAP, PCWP should all come together and reach a common number as the tamponade takes effect and actual flow gradients cease to exist. The PAP of 42/28 while both systolic and diastolic are elevated, does not suggest that a tamponade is present in and of itself. The PCWP, being a single number or mean, doesn't afford us anything to compare against to say there isn't a tamponade possibly present. We did not include

A, F, J or R because while these commonly reflect an elevated PAP, the PCWP is more indicative of LVEDP further downstream. Granted the PCWP can be skewed sometimes by these conditions, the textbook answer would not include them however.

33. d. A, F, J, R This option presents with a low-normal PCWP and a significantly elevated PA pressure. This is suggestive that the etiology is limited to the pulmonary system and does not likely involve the left side of the heart downstream. When considering the option of a PE it would likely be assumed that the PE must be in an area of the lung that is not affecting the flow from the catheter tip location (hence the normal PCWP) but be significant enough to cause measurable pulmonary vasoconstriction (elevated PAP)

33. e. A, B, C, D, E, F, G, H, J, M, N, P, R The enormous number of possibilities to cause an elevated CVP illustrates how limited the data gleaned from a CVP catheter is. Basically, *anything* going wrong downstream can cause an elevated CVP.

33.f. C, E An elevated CVP with a low PAP presents a very interesting scenario. Apparently there is high preload to the RV but the pulmonary artery isn't seeing any of it. Of the choices offered, an 'obstruction' must be occurring between the CVP measurement point (the opening of the vena cava to the RA) and the PAP measurement point in the distal pulmonary trunk. Causes could be a tricuspid valve that won't open or close, an RV that won't pump or a pulmonic valve that won't open. Only two are offered in the set of possible causes.

33.g. P While many things will cause an elevated CVP, PA systolic and PCWP (wedge pressure), only one thing elevates all of them to roughly the same value..... pericardial tamponade. Look for these numbers in scenario based questions where the coronary arteries may have been damaged (PTCA) or the heart punctured (trauma, transvenous pacer placement, etc.)

TEST TIP
All hemodynamic parameters converging on the same value ⇒ pericardial tamponade

33.h. I, K, L, O, Q All of the values given are grossly low. This patient either has too big a vascular capacitance, (pipes are too big) or inadequate volume (not enough fluid to fill the pipes).

33.i. K These values signal inadequate preload to the right side of the heart, vasodilation (or lack of vasoconstriction) and increased cardiac output. The only scenario that would likely present this combination is the patient in early or 'warm stage' sepsis. In early sepsis the bacterial invasion causes the release of bacterial endotoxins of which bradykinin is one. These endotoxins promote vigorous inotropicity and chronotropicity initially while inhibiting vasoconstriction. As such we see values that appear 'shocky' but a substantial CI may be present. The elevated CI will fade when the system can no longer perform at this hyperdynamic level and exhaustion sets in. ARDS would not be a selection here because ARDS by itself (and we must consider each choice independently) would not typically promote an elevated CI, it may or may not precipitate the low SVR. Remember, if the body needs to vasoconstrict and it can.....*it will*! A low SVR means the body can't vasoconstrict, this almost always points at some form of redistributive shock state. So why is neurogenic shock not an option here? Again, the CI. The hallmark of neurogenic shock is a shocky patient *without* tachycardia or rate compensation.

TEST TIP
"shocky" hemodynamics <u>without</u> tachycardia ⇒ neurogenic shock

33.j. I These values present the ideal picture of the patient in hypovolemic shock. Low CVP identifies no obstructions downstream and either too much capacitance (vasodilation) or inadequate volume (hypovolemia). The elevated SVR follows the rationale offered in the previous question. When the body needs to vasoconstrict, and it can.....*it will*! This patient's vascular system is constricting down in an attempt to maintain pressure. The CI has fallen to a low value. Without adequate preload, you cannot have adequate cardiac output. Remember CO=SV X HR, no SV equals no CO.

33.k. A, B, C, D, E, F, G, H, J, M, P, R The values offered signal excessive preload pressures at the CVP/RAP. This helps to rule out causes such as hemorrhage or vasodilator excess. The elevated SVR tells us the patient is vasoconstricting, the question to ask is, "why?" Answer, the tissues are identifying inadequate oxygenation (perfusion) and they're attempting to compensate for some form of shock. The fact they are vasoconstricting rules out forms of distributive shock (anaphylactic, neurogenic & septic). Finally the CI tells the real story. We have preload, and we have functioning 'pipes'. The problem is in the pump. Somewhere between the CVP and systemic arteries lies a failure or obstruction. Obviously with additional data, such as PAP or a PCWP you could narrow the possibilities further. Work through some of these in your head. By now you should be getting very comfortable troubleshooting this system.

33.I. A, C, E, F, J, R This patient data sample uses the exact same values as the previous but adds one more data point, a low PCWP or wedge pressure. This tells us to start with the same causes simply ruling out anything past the pulmonary system. The low PCWP suggests that the obstruction is occurring before the left atria most likely. In fact with this one extra data point, we are able to reduce the possible causes by half.

34. B The waveform description should lead you to suspect either a PCWP (wedge) waveform or a CVP. However, for the catheter to have pulled back to the CVP you would anticipate seeing the sharp RV pattern first as well as typically requiring quite a bit of displacement. A new catheter placed in a proper PA position may inadvertently migrate deep just enough to wedge simply as a result of the catheter warming to body temperature and becoming more flexible allowing a significant curve (if present) in the RV to straighten somewhat thereby pushing the tip more distally.

35. A The trick here is in the wording. The catheter balloon is "designed" to hold 1.5ml. How much should you inflate with when the catheter is in the patient? Only enough air to cause a waveform change, with a maximum limit of 1.5ml.

TEST TIP
When wedging a PA catheter, you only inflate the balloon with enough air to cause a change in waveform and <u>never</u> more than 1.5ml total.

36. C When the stiff catheter is initially placed a large loop may exist as it passes through the RV. As the catheter warms to body temp and becomes more pliable, the slack created by the loop through the RV may allow slight migration deeper into the PA with inadvertent wedging. Position changes can cause inadvertent wedging as well. Frequently having the patient move or cough can displace the inadvertent wedge back to an acceptable PA waveform position.

37. A Air trapped in the tubing will absorb energy or pressure waves from the artery before reaching the transducer, this is increased with a rise in altitude or drop in atmospheric pressure as the bubble becomes larger following Boyle's Law. Pressure waves will compress the air bubble and thus the air bubble will rebound during pressure drops to cause reduced or over damped waveforms. Kinking of the pressure tubing or closing via a stopcock will also limit energy or pressure waves from reaching the transducer. The best option, option A, would cause a reduction in the size of any air bubbles in the system and would therefore take the system to a more under-damped state.

38. D The left anterior fascicle and the right bundle branch both descend from the AV node towards the apex via the anterior ⅔rd's of the septum. Ischemia of this region not only affects muscle but also will affect the blood supply to these conduction pathways. "Flash" pulmonary edema is associated with acute mitral valve insufficiency. The papillary muscles, which assist the mitral valve, are supplied by the same coronary system supplying the anterior wall; hence an acute anterior wall MI may cause papillary muscle ischemia or rupture. The AV node is supplied by the RCA in the majority of the population (~90%) so an isolated anteroseptal infarct (LAD lesion) shouldn't affect the AV node much if at all.

39. D Pressure line readings should always be assessed at end-exhalation, regardless of ventilation mechanism. The baseline wander typical of hemodynamic pressure lines will reverse axis when moving from spontaneous to mechanically ventilated patients but the end exhalation point in the baseline wander will be the same.

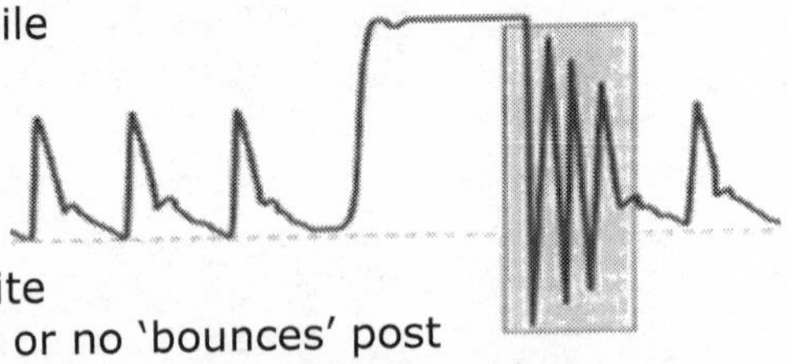

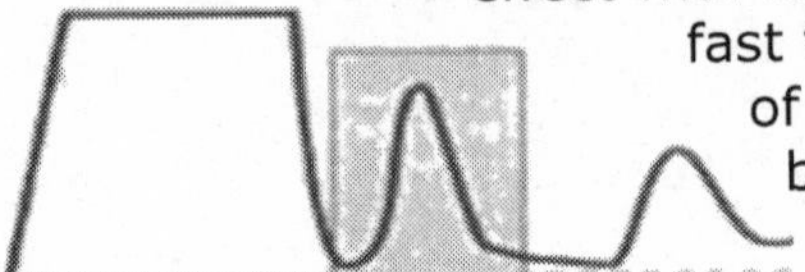

40. C The "fast flush" described indicates the system is too dynamic. The most likely cause would be excessive tubing, incorrect tubing type, very distal placement or low pressure in the pressure bag (which may occur while descending to lower altitudes) and finally low pressure bag fluid level. This is referred to as under-damping (see right). Over-damping would be the opposite effect with little or no 'bounces' post fast flush (see left). The typical causes of over-damping would be air bubbles, closed stopcock, kinked line or an over-pressurized pressure bag (which may occur while ascending to altitude). The term 'whipping' is applied in reference to invasive pressure monitoring in a couple applications. First 'whipping' has been applied to infer the exaggerated waveform readings that may occur in very distal peripheral arterial line insertions. This effect is caused by the inherent dynamic properties of the patient's extended vascular system between the source (heart) and the measurement point being the insertion site of the arterial line. Stiff tubing is utilized in the invasive monitoring set to minimize this effect by the tubing itself. However very distal insertion sites have more expandable "tubing" (blood vessel) between the catheter tip and the source of the pressure waveform (the heart). "Cushioning" is an attractive distracter, something I prefer to sit on.

TEST TIP
The most common cause for over damping is "air in the system"

41. D For the Intra-Aortic Balloon Pump (IABP) to be effective, the aorta must remain intact and the aortic valve competent. Failure of either of these will allow the IABP to cause harm. While peripheral vascular disease (PVD) is commonly listed as a contraindication to IABP use, the fact is anyone with enough CAD to require an IABP most likely has some degree of PVD. As such, the presence of PVD is typically reviewed as a 'relative' contraindication.

42. A The tip of the IABP balloon/catheter should typically rest 2-4cm below the aortic arch. Higher placement will promote occlusion of the left subclavian artery. Lower placement of the balloon can compromise perfusion of the renal arteries. Option D is an attractive distracter, one would typically place an invasive pressure transducer at this location.

43. D Once blood enters the dry, cold, helium rich environment of the IABP balloon, it quickly dries, flakes and changes to a rust or brownish color. All other options offered are attractive distracters. When an answer sounds "too exotic" or "too elaborate", it probably is. Helium is highly insoluble in blood and will thus form a large air embolus if the pump continues to operate. Obviously stop the IABP and immediately place the patient in trendelenburg position to keep the embolus trapped in a distal extremity. Route the patient to a cathlab immediately.

TEST TIP
"Rust colored flakes in the IABP tubing" ⇒ IABP balloon rupture

44. D These two core concepts are at the heart of IABP function. Like many things in life, timing is everything. Your goals: improve coronary perfusion; take the work off the ventricle.

45. B The IABP has been used for several pathologies, including as a bridge device between PTCA and CABG. ECMO is a bypassing of the systems, which the IABP is designed to assist, thus the IABP's use would be moot. An incompetent or diseased aortic valve is a specific contraindication to the use of the IABP.

46. B The device must be placed in a mode other than 1:1 assist when timing. The first number in this ratio is always 1. The second number represents which heart beat will be assisted before the cycle repeats. Hence in 1:1, every heartbeat is assisted by the IABP. In 1:2 mode, every other beat would be assisted, 1:3 mode, every third beat, etc. Options C and D are distracters that play on test takers unfamiliar with IABP assist ratios. When timing the IABP one must compare an assisted beat to an unassisted beat, hence 1:1 will not allow the proper comparison.

47. D Careful monitoring of the SpO_2 on the left hand specifically is warranted to alert you to deep migration of the IABP and blockage of the left subclavian artery. Specific SpO_2 monitoring of the insertion side lower extremity alerts you to thrombus formation at the insertion site and compromise of that extremity's blood supply. Renal output monitoring should alert you to shallow positioning or withdrawal of the IABP balloon and obstruction of the renal arteries.

48. C The IABP has "mechanically failed", in other words, "broke, 10-7, out of service". Setting the IABP to pressure mode or altering it's triggers will not fix the problem. In this case the balloon's stasis becomes the problem with blood beginning to clot on the balloon itself. Manual inflation of the balloon with the proper size syringe (supplied with the balloon from the manufacturer) every 30 minutes is warranted. Otherwise the balloon will at a minimum shower the insertion side leg with emboli upon removal.

49. B The inflation waveform is indicating that the dicrotic notch formed doesn't reach the same level as the unassisted beat's dicrotic notch. The sharp 'V' created is too early. The deflation wave appears properly timed forming a nice deep 'U' shape. Early inflation would indicate the balloon is filling and obstructing the aorta before the end of systole has occurred. This is a potentially harmful timing error and needs immediate correction. Compare this timing strip and others in the self-test to the strip below, which has good timing. (Note the assist mode below is 1:2)

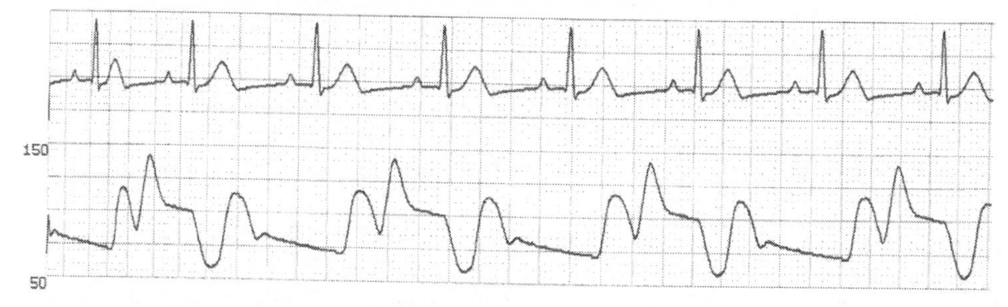

50. A The beginning of the inflation upstroke is too late. You can see the dicrotic notch of the assisted arterial waveform clearly visible before the inflation upstroke. Deflation looks good.

51. D This strip is demonstrating early deflation. The nice deep 'U' created by proper deflation is not present. One can clearly see the lowest point of the deflation downstroke, which then gradually rises to the patient's unassisted diastolic pressure prior to the next unassisted arterial waveform. The downstroke of the balloon deflation should blend seamlessly into the next arterial waveform's upstroke forming a nice deep 'U' shape.

52. B This ECG shows distinctive ST elevation patterns in V1, V2, V3 & V4 clearly identifying an anteroseptal wall AMI. There appears to be slight ST elevation in leads I and aVL as well. This would suggest probable lateral wall AMI involvement or 'extension' as well. Answer B is selected because it identifies more ischemic area than answer C. In reality the diagnosis of anteroseptal wall AMI with lateral extension would be more ideal. The ST depression noted in II, III and aVF would be considered reciprocal findings of the ST elevation in the precordial leads. Remember the key to identifying an AMI on the 12-lead ECG is ST elevation.

53. C The best choice is the proximal LAD. The LAD provides the majority of flow to the anterior wall, anterior $\frac{2}{3}^{rd}$'s of the septal wall and the diagonal branch of the LAD feeds portions of the anterolateral wall. An argument could be easily made that this may be an occlusion of the left main or LCA, however that option is not offered as an answer option. An RCA lesion would be expected to affect the RV, left posterior and inferior walls and/or the AV node. A distal LAD lesion would not typically cause the widespread injury pattern demonstrated here.

54. D As discussed in prior question rationales, an anterior wall AMI may precipitate mitral valve papillary muscle failure or rupture. Acute flash edema will result as the incompetent mitral valve allows high pressure from the left ventricle to preferentially flow backwards into the pulmonary circulation.

55. D The ST elevation in II, III and aVF is quite obvious, identifying the inferior wall portion of the AMI. The ST depression in V1, V2, V3 & V4 are reciprocal findings of posterior wall ischemia. When a mirror-test is performed on this ECG the posterior leads clearly demonstrate ST elevation and pathological Q-wave formation (suggesting infarcted tissue). Physiologically this makes sense as the inferior and posterior walls are supplied by the same primary supply, the RCA. The lateral extension is not uncommon. The lateral wall shares larger and larger amounts of its blood supply between the LCA and RCA as the patient ages and develops increased CAD due to increased collateral development.

56. A Patients with inferoposterior wall AMI typically have a lesion/obstruction of flow in the RCA. If the lesion is proximal enough the RV may be ischemic as well. Because RVI doesn't lend itself to easy identification on the standard left sided 12-lead ECG, it may not be obvious. RVI causes a drop in pre-load, which may be exacerbated by NTG. These patients require preload augmentation along with reperfusion therapy.

57. D The question stem is full of distracting information. In timed exams you might consider reading the answer choices prior to reading the question, this is especially beneficial when you see an entire paragraph is used for the question stem. This allows you to have an idea what you're looking for prior to spending large amounts of valuable time sifting through unnecessary data.

Remember hydrostatic pressure is effectively the pressure created by the system itself, the pump, the tension of the pipes and the amount of fluid in the pipes. Hydrostatic pressure is commonly used synonymously with 'blood pressure'. Osmotic pressure is the pressure created by the number of osmotically active particles, or "the amount of stuff" in the fluid referenced.

Generally speaking hydrostatic pressure pushes fluid out of blood vessels, osmotic pressure should pull fluid back in (relative to surrounding compartments.) Of the data offered in the question, the RAP of 9mmHg is high, (remember the RAP and CVP are the same thing). A high CVP suggests high pressure in the central circulation such as the vena cava.

58. B End stage emphysema and cor pulmonale commonly occur together. Remember cor pulmonale is a hypertrophied right ventricle secondary to high pulmonary pressures. The chronically hypoxic patient will have chronically vasoconstricted pulmonary vasculature. This causes a volume overload as fluid backs up to the RV. As the RV is overloaded, fluid will back up to the RA and may stretch the RA as well. As the RA is stretched this promotes "surface arrhythmias" such as atrial fibrillation. Emphysemics are commonly referred to as "pink-puffers". The 'puffing' label comes from the repeated pursed lip breathing the emphysemic patient employs to maintain adequate pressures to prevent collapse in the small airways during exhalation. The 'pink' label comes from a pinkish tone to the skin as the patient becomes polycythemic. Building RBC's ultimately causes hemoconcentration reflected in an abnormal H&H as well as promotes increased 2-3DPG levels.

59. A All forms of shock ultimately result in reduced cardiac output. However, in the early stages of septic shock, endotoxins released from the invading bacteria cause increased inotropicity and chronotropicity. As such a hyperdynamic cardiac output may be seen.

60. D Ultimately, true shock is a result of anaerobic metabolism. The primary waste product associated with anaerobic metabolism is lactic acid, which is directly measured with a lactate level. The presence of a low PaO_2, high $PaCO_2$ or elevated K^+ may be common with various forms of shock but their presence does not determine shock in and of itself. Normally lactate levels run 0.7-2.1mEq/L (1-2mEq/L for ease of recall). A lactate level greater than 4mEq/L is commonly considered a significant finding.

61. D Coronary blood flow is dependent upon all the factors listed with coronary perfusion pressure being determined by the difference between the aortic diastolic pressure and left ventricular end diastolic pressure (AoDP-LVEDP). Additional factors influencing coronary blood flow include presence and severity of transluminal obstructions and coronary vascular tone.

TEST TIP
Coronary perfusion pressure (CoPP) = AoDP - LVEDP

62. D Most of us have been taught that the heart is perfused during diastole. Unfortunately this is because many authors/ educators only consider the left heart when teaching this principle and the left ventricle is perfused almost entirely during diastole. However the right ventricle is perfused predominantly during systole to include the SA and AV nodes. Should a question such as this arise on the exam and the question's stem does not specify left ventricle vs. right, "diastole" is the most likely answer they have keyed while not completely accurate. (Don't beat yourself up if you answered B. Don't beat me up either. I'm delicate..... like a flower.)

63. D This should be a simple recall answer when you take the test(s). Be cautious to have a good understanding of the basic formulas and equations for this section of the exam. Stroke volume, the amount ejected with each systolic contraction of the heart multiplied by the rate of the heart will provide cardiac output.

64. B While SVR and afterload can be used synonymously in many instances and some may use EF and contractility somewhat synonymously, none of these options are entirely correct. Heart rate, while factorial in cardiac output, is not considered factorial with stroke volume.

65. A Since the left ventricle is perfused during diastole, diastolic pressure is obviously factorial as is the time that the diastolic pressure has to perfuse the ventricle itself. Fast heart rates equate to shorter diastolic times, slower heart rates equal longer diastolic times. This is part of the rationale behind the use of beta-blocking drugs during AMI management. PA systolic is a "left-field" answer and hopefully you tossed that one out relatively quickly. SVR does affect the left ventricle's oxygen demands but not supply. "Oxygen demand" is a very attractive distracter since the term 'supply and demand' is so commonly used in medicine.

66. C Myocardial wall tension is the primary determinant of myocardial oxygen consumption with contractility being a close second. Heart rate and BP provide good indicators of that consumption but are not determinant per se. Remember wall tension can be measured using the **Law of LaPlace;**

$$\mathbf{T(tension)} = \frac{\mathbf{P(pressure) \times r(radius\ of\ the\ container)}}{\mathbf{2 \times h(thickness\ of\ the\ container)}}$$

Per the formula's application to the heart, ventricular wall tension is directly proportional to intraventricular pressure and radius of the ventricle; while inversely proportional to the thickness of the myocardial wall. Hence wall tension and myocardial oxygen consumption is high in the left ventricular hypertrophy patient as the ventricle has a larger diameter and thinner wall.

67. B The *SHOCK study* (2000) determined that the use of IABP's for AMI's with cardiogenic shock was helpful and thus indicated in any case of AMI with cardiogenic shock. Hopefully you identified the use of 'all' in answer C and eliminated that from your likely choices along with D. A is a reasonable choice but not a better selection than B.

68. A All of the choices are complications associated with the use of an IABP, distal limb ischemia is the most common however. Teams responsible for transporting IABP's on a regular basis typically carry two pulse oximeters. One is placed on the left hand to identify if the balloon tip migrates too deep blocking the left subclavian artery; the other is placed on the foot of the limb with the IABP insertion to identify distal limb ischemia.

69. D Early inflation and late deflation are the two harmful timing errors with late deflation being most harmful. Keep in mind that with late deflation, the balloon is up in diastole, systole begins with a full ventricle trying to squeeze with it's highest pressures against the inflated balloon, then the balloon 'drops'. This has the potential of causing a multitude of high-pressure ruptures in the brain and coronaries, not to mention the enormous oxygen demand increase placed on the heart as pressure (refer to the Law of LaPlace in question 66's rationale) increases.

70. C The 'trigger' is what tells the pump when to perform the cycle of inflating and deflating. The primary trigger is the ECG with a secondary or backup trigger of the arterial pressure. Do not confuse the 'trigger' with 'timing'. You will always use the arterial waveform to set *timing* but the ECG is your first choice for *triggering*.

71. B Remember flight physiology. "Boyle's balloon" sound familiar? If not you should probably review that section.

72. D When the balloon pump purges it stops pumping to evacuate the helium from the system and reload the appropriate volume based on atmospheric conditions at that time. If the patient is very dependent on the IABP's assist, he/she may decompensate very rapidly. You need to be prepared to assist with fluid boluses and/or vasopressors.

73. D When IABP patients are in an erratic rhythm, especially without coordinated atrial activity, the ECG trigger may have difficulty anticipating the QRS. Setting the IABP to a "pressure" or "internal trigger" allows it in many cases to coordinate timing better with the erratic rhythm of the heart.

74. D You must know the anatomy of the femoral vein for purposes of these exams. (Refer to the diagram below.) Cannulation of the femoral vein is a common skill set found in aeromedical transport.

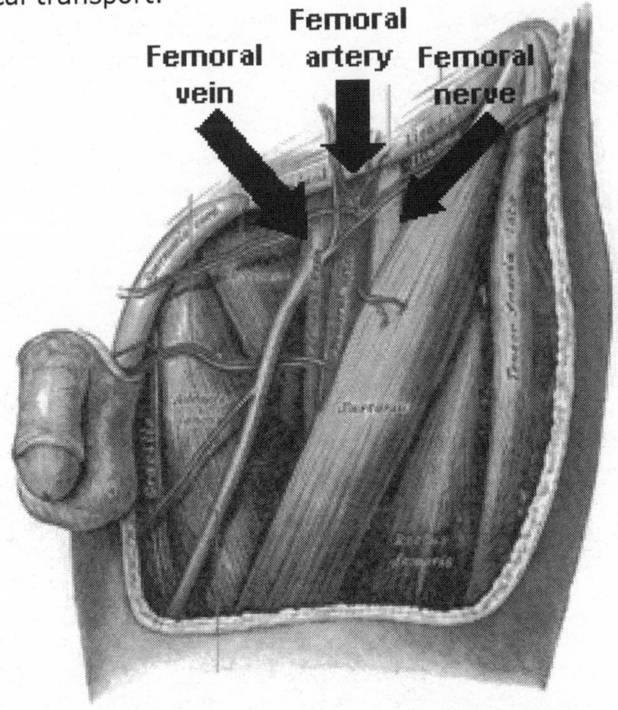

modified from http://us.i1.yimg.com/us.yimg.com/i/edu/ref/ga/l/549.gif
"Gray's Anatomy" -- Copyright 1918

75. B While all of the options presented are complications of subclavian cannulation, the most significant or acutely dangerous is pneumothorax. An AP chest X-ray is indicated with every subclavian access for this reason.

76. B Cannulation of the right IJ vein is technically easier and a safer procedure than subclavian vein cannulation, but not commonly practiced in EMS. The femoral vein is technically the easiest access (with adequate blood pressure for palpation of landmarks) but carries with it the possibility of inaccuracy if the catheter tip is not above the diaphragm and frequent inadvertent arterial cannulations in the hypotensive patient. A right subclavian carries the potential of a pneumothorax on the right and dropping both lungs is a 'bad idea'. The left subclavian is a 'freebie'. If you drop the lung there's already a chest tube in place. (Even if you opted to perform an IJ as a technically easier procedure you would want to place it on the left to minimize risk of dropping the other lung.)

77. C When a PA catheter is inserted via the IJ, the various compartments are encountered 'on the tens'. CVP/RAP at 20cm, RVP at 30cm, PAP at 40cm, PCWP at 50cm. If the catheter is inserted via a subclavian approach, add 5cm to each of these. If the catheter is inserted via a femoral approach, add 20cm to each of these.

78. D Most cardiovascular interventional procedures involve the administration of large amounts of heparin to prevent blood-clotting in/on the equipment used during the procedure. Reversal of this heparin is required at the end of the procedure using protamine sulfate.

79. A Protamine sulfate is obtained from fish sperm or recombinant DNA technology. It commonly results in significant histamine releases, especially in patients with multiple allergies or known allergies to fish. Typically a small 'test dose' of protamine is administered to see how the patient will respond first. You should anticipate flushing, hypotension, tachycardia, bronchospasm, pulmonary hypertension and ultimately anaphylaxis.

80. B Normal EF is 60-65%, some texts will suggest 60% +/- 7%. Once the EF drops below 40% the patient is considered 'high risk' for invasive procedures with or without concomitant disease.

81. C "Mobitz I", "Wenckebach", "2nd degree type I", whatever you want to call it, that's what it is. You should note the PRI getting longer with each captured QRS until finally there is a dropped QRS.

82. C The inverted P waves tell you they are retrograde conducted along with the very short PRI. The ventricular rate is less than 60 which coincides with a nodal pacemaker's intrinsic rate (40-60/min) and not an 'accelerated' junctional rhythm.

83. B The overall rhythm is sinus in origin but there is a significantly wide QRS clearly visible in II, aVR and V1. The QRS width exceeds 120ms thus a bundle branch block (BBB) is present. Using V1 you can quickly ascertain right vs. left BBB and in this case it is a right BBB. Finally, the negative QRS complexes in II and III identify a pathological left axis deviation, you can also identify this by lead I's overall positive deflection and aVF's negative deflection. If so inclined you could even use a hexagonal system and identify that the actual axis is approximately -85°. Regardless of how you identify the pathological left axis deviation this should lead you to identify a left anterior fascicular hemiblock (LAFHB). Right BBB plus LAFHB equals a "bifascicular hemiblock". Now, if all this axis and fascicular talk has your head spinning, try attacking this from an elimination perspective. Is it A, NSR with 1st degree AVB? No the PRI is less than 200ms. Is it C, accelerated IVR with RBBB? No, first of all there are clearly P waves correlating with the QRS's therefore it's not idioventricular and furthermore, you can't identify a BBB in a ventricular rhythm because the conduction pathways are completely aberrant. Finally is it D, sinus tach with LBBB. The rate is clearly less than 100/min, thus it cannot be tachycardia. So you see, even if you're not 'up to speed' on detailed electrophysiology, this test question can be overcome.

84. C The ventricular irritability being seen is typical of a PA catheter tip that has pulled shallow and is now banging around in the ventricle causing arrhythmias. Assuming the catheter was inadvertently withdrawn some, you simply need to verify the balloon is down and withdraw it to a CVP position for optimal safety, note the depth and re-secure the catheter.

85. D The arrows along the bottom of the strip identify when the pacer triggered. The pacer spikes above those do not demonstrate an obvious capture waveform immediately following. The large complexes with a dot at the top of the QRS are the patient's intrinsic beats and the 'dots' are the monitors 'flags' identifying the complex's r-wave. What is the key to identifying electrical capture after a pacer spike? You should see a wide complex with bizarre morphology similar to a PVC but even *more important* than that, you should see a T-wave immediately following that complex with an inverse polarity. Remember, if your pacer depolarized the myocardium, you will see repolarization immediately following.

86. D If you just glanced at the rhythm you probably identified it as ventricular fibrillation (VF) or possibly ventricular tachycardia (VT). On closer inspection you should be able to identify the rhythmic oscillation of the 'fib-like' pattern. This is torsades de pointes, also known as multi-focal ventricular tachycardia. This particular rhythm should be treated rapidly with magnesium.

87. B You should have determined that the PA catheter has spontaneously or inadvertently 'wedged' and you are seeing a PCWP waveform. You need to verify the balloon didn't somehow become inflated and once that is assured ask the patient to cough forcefully or even perform a valsalva maneuver to see if the change in pressures can un-wedge the catheter. Moving the patient from side to side can be helpful in this scenario as well. If these maneuvers do not correct the catheter placement, then the catheter needs to be pulled back slightly until a PA waveform reappears.

TEST TIP
Inadvertent wedge ⇒ verify balloon down, have patient cough & change position, then withdraw the catheter only until the PA waveform reappears

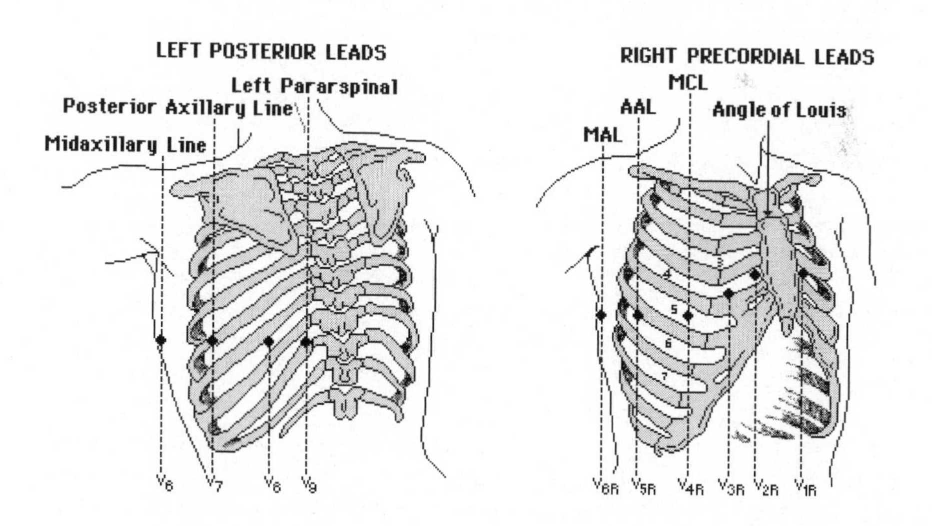

88. B The invasive pressure waveform (seen below) begins with an obvious sharp waveform, which includes notching on the left side (indicated by arrow A) or the ascending side (an anacrotic notch). This should identify an RV waveform to you. After the sixth wave in the tracing (letter C), the lowest point of the deflections, or the baseline, shifts upward from the previous ~10mm Hg (see arrow B) to ~18-30 (see arrow C). This signifies that the catheter just transitioned past the pulmonic valve, which is now functioning to maintain the diastolic pressure (the lowest point in the deflection), higher than that found in the RV. You should also note that at the point which the baseline shifts up (letter C), the anacrotic notch and atrial kick disappear and a more defined dicrotic notch appears indicating closure of the pulmonic valve.

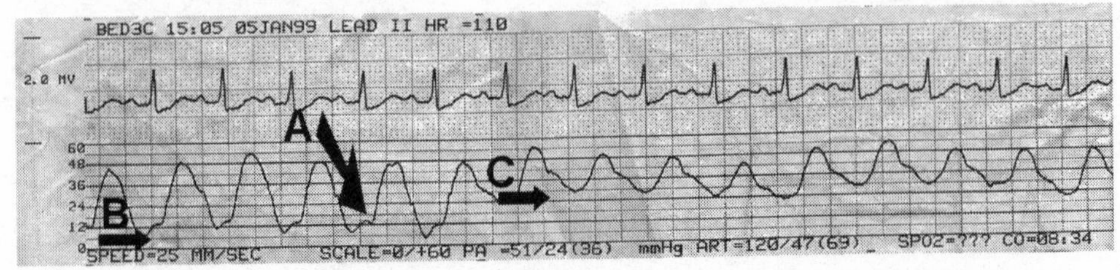

89. B The notch (indicated by arrow A in the strip above) is the end of the atrial kick, which is creating the upward deflection just prior to the notch. Recall that atrial kick provides a substantial part of ventricular filling.

90. A Again referencing the strip above, follow the Swan tracing along the level of arrow C. You will see that the actual lowest points on the waveforms dip into the 18-24mmHg range some places and only down to 24-30mmHg range in others. This natural baseline wander is typical of breathing and the primary reason for measuring all hemodynamic pressures at end exhalation to maintain consistency and accuracy.

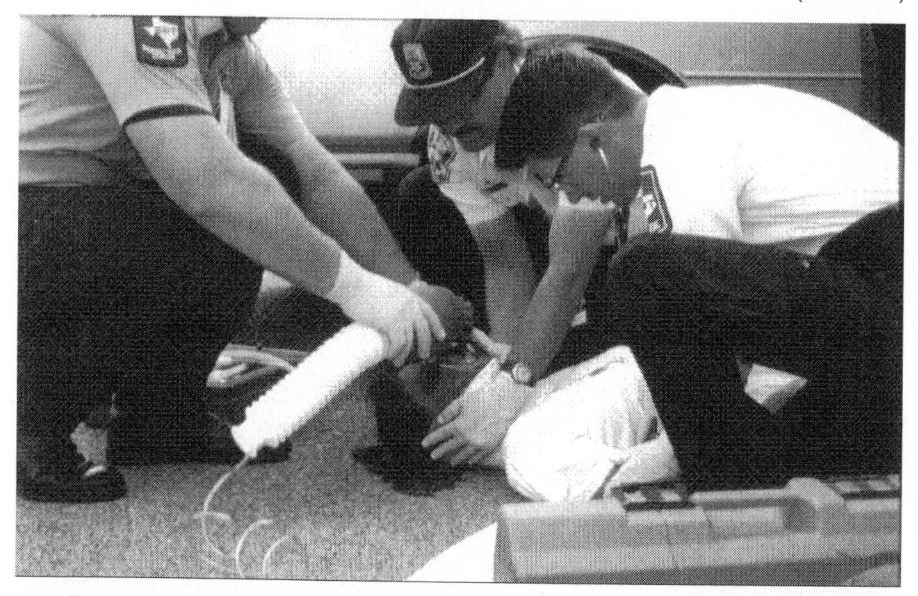

Wingfield WE; *Blind Nasal Intubation of Maxillo Facial Trauma*, Emerg, Jul 1997, pp14-16

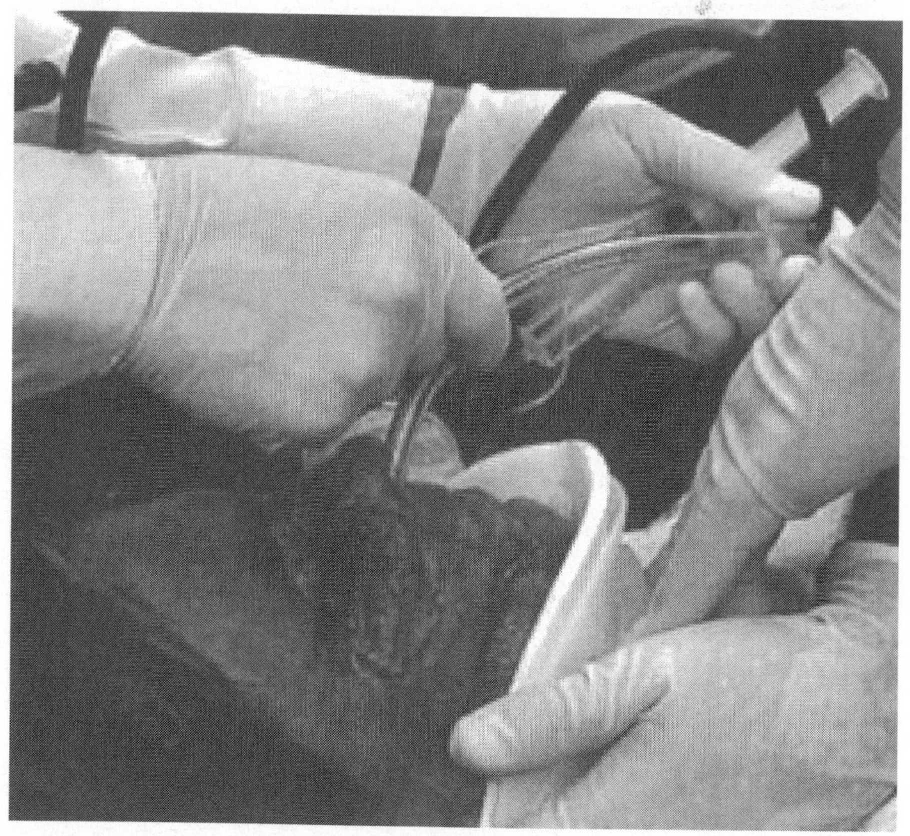

AIRWAY MANAGEMENT & PHARMACOLOGY

1. "A" comes;
   a. before B and C
   b. before B after C
   c. after B before C
   d. after B and C

2. Intubation is clinically indicated in which of the following patients?
   a. 38yo male pulled from a house fire, he is breathing 20 breaths per minute with good chest excursion, he is unconscious with 80% BSA burns including his face and neck.
   b. 67yo male presenting to the ED with chest pain, breathing 40 breaths per minute and dyspneic. SaO_2 is 65%, PaO_2 is 44mm Hg, he had no prior pulmonary pathology.
   c. 19yo female presents in an area known to be frequented by heroin users. She has pinpoint pupils, no obvious trauma and is breathing at four breaths per minute.
   d. all of the above

3. A patient that is unconscious and unresponsive or near death would be managed most appropriately using the;
   a. crash airway algorithm
   b. difficult airway algorithm
   c. RSI airway algorithm
   d. failed airway algorithm

4. You began managing your patient utilizing the difficult airway algorithm but the patient suddenly deteriorates to an SaO_2 maintainable at 80% and she has lost consciousness. You should now begin managing the patient's airway utilizing the;
   a. crash airway algorithm
   b. continue with the difficult airway algorithm
   c. RSI airway algorithm
   d. failed airway algorithm

5. Which of the following patients should not be managed using the "Failed Airway algorithm"?
    a. the patient has an SpO_2 of 85% with BVM assist at an FiO_2 of 1.0
    b. the patient has endured 3 failed intubation attempts
    c. you have just found the patient, he is unconscious, unresponsive and near death
    d. all of the above would be managed using the Failed Airway algorithm

6. Which three criteria must your patient possess to appropriately indicate RSI as the airway technique of choice?
    a. they require sedation, paralyzation and intubation
    b. they require sedation, paralyzation and you can ventilate them with a mask
    c. they require sedation, paralyzation and you believe you can intubate them
    d. they require sedation, you can ventilate them with a mask and you believe you can intubate them

7. Assessment of BVM success is best obtained via;
    a. SpO_2
    b. fogging of the BVM mask
    c. bag compliance
    d. direct visualization of chest rise

8. Scoring techniques used for identification of the difficult airway prior to intubation include all but which of the following;
    a. Modified Mallampati Test (MMT)
    b. Upper Lip Bite Test (ULBT)
    c. Cormack-Lahene Grading System
    d. 3-3-2-1 External Airway Assessment System

9. Which of the following clinical signs would not suggest the probability of a difficult intubation.
    a. obesity
    b. protruding lower incisors
    c. spinal immobilization
    d. Mallampati Score of III

10. You have encountered a 200lb male patient requiring emergent intubation. He is initially cooperative and you perform an external exam of the airway. You assess the following; MMT – IV, ULBT – Class I, 3-3-2-1, secretions controlled and/or manageable, neck range of motion (ROM) is full w/o pain/restrictions, dentition is good. SaO_2 is 93% on a simple mask at 6lpm, he is clean shaven and without trauma or deformity of the face and jaw. Which of the following is the best airway management plan?
    a. Laryngoscope with blade of choice, 8.0 ETT, intubate using a standard RSI technique, have assistance close by.
    b. Laryngoscope with blade of choice, be ready to quickly switch to Miller blade if visualization is poor, 8.0 ETT, utilize an RSI technique but be prepared to move to a rescue airway if needed, have an LMA/CombiTube prepared at bedside and open as well as a surgical airway kit.
    c. This patient is likely not intubatable via standard techniques, ask for assistance for fiber-optic placement or surgical airway placement immediately.
    d. Use an awake technique to assess the airway more thoroughly. Using a Macintosh blade provides optimal visualization. Have assistance close by and plan accordingly for a failed intubation.

11. Which of the following is not intended to improve glottic visualization?
    a. Sellick's Maneuver
    b. External laryngeal manipulation (BURP maneuver)
    c. head elevation to sniffing position
    d. all of the above are intended to improve glottic visualization

12. The primary rationale for the use of RSI is;
    a. to provide intubation conditions in emergent airway management
    b. to prevent vomiting
    c. to prevent movement
    d. to provide amnesia

13. When utilizing an "awake technique" during the difficult airway algorithm, the best drug for induction from the choices below would be;
    a. midazolam (*Versed®*)
    b. diazepam (*Valium®*)
    c. ketamine (*Ketalar®*)
    d. etomidate (*Amidate®*)

14. Per AHA ACLS guidelines, all endotracheal tube placements should be verified by a "primary" and "secondary" verification technique. Which of the following is the most reliable "secondary" verification?
    a. visualizing the tube passing into the trachea
    b. esophageal bulb detector
    c. end-tidal CO_2 detector (i.e. - *Easy Cap®*)
    d. capnography

15. The purpose of pre-oxygenating the RSI candidate is to;
    a. replace the nitrogen in the functional residual capacity (FRC) with oxygen
    b. optimize the SaO_2 and PaO_2 prior to intubation
    c. relax the pulmonary vasculature allowing for optimal pulmonary blood flow
    d. all of the above

16. The best way to pre-oxygenate a spontaneously breathing patient prior to RSI would be;
    a. Apply a non-rebreather mask (NRBM) with flow at 10-15 lpm.
    b. Apply a bag-valve mask allowing the patient to breath passively through it with flow at 10-15 lpm.
    c. Apply a bag-valve mask ventilating with the patient with a flow of 10-15 lpm.
    d. All of the above are equally efficacious to pre-oxygenate

17. The recommended time to pre-oxygenate a patient properly prior to RSI is;
    a. 2 minutes
    b. 3 minutes
    c. 5 minutes
    d. 8 minutes

18. Use of supra-glottic rescue airway devices is recommended in which of the following airway algorithms?
    a. crash airway
    b. difficult airway
    c. RSI airway
    d. failed airway

19. Which of the following patients will desaturate the fastest after an appropriate pre-oxygenation (de-nitrogenation)?
    a. 67kg, 44in tall, 8 year old male with a fractured femur
    b. 127kg, 68in tall, 34 year old male with acute appendicitis
    c. 65kg, 62 in tall, 84 year old female with a hemorrhagic CVA
    d. 14kg, 29in tall, 3 year old with status seizures

20. The most common cause of anaphylactic reactions would come from which of the following?
    a. lidocaine
    b. neuromuscular blocking agents
    c. morphine
    d. latex

21. The purpose of administering opioids prior to intubation is to;
    a. intensify the sedation so the patient doesn't remember the event
    b. intensify the relaxation so the patient presents with optimal intubation conditions
    c. improve hemodynamic supply to the lungs to optimize oxygenation
    d. attenuate the sympathetic response associated with laryngoscopy

22. Pre-medication of pediatric patients with atropine provides which of the following?
    a. prevents tachycardia associated with succinylcholine use
    b. prevents hypersalivation associated with ketamine use
    c. prevents bradycardia associated with ketamine use
    d. prevents tachycardia associated with instrumentation of the airway

23. Defasciculation prior to administration of succinylcholine will prevent;
    a. myalgia
    b. elevated potassium level
    c. malignant hyperthermia
    d. elevated intra-cranial pressures (ICP) during laryngoscopy

24. Fentanyl, when given in large doses in a rapid fashion, may precipitate all but which of the following?
    a. coughing
    b. "rigid-chest syndrome"
    c. sudden respiratory arrest
    d. retrograde amnesia

25. Which of the following is not characteristic of etomidate (*Amidate®*)?
    a. occasional myoclonus may be seen
    b. duration of action is ~100sec per 0.1mg/kg administered
    c. very hemodynamically stable
    d. adrenal hypersecretion has been reported in the literature

26. Succinylcholine (SUX) is contraindicated in all but which of the following?
    a. Duchenne's
    b. Previous history of sibling with malignant hyperthermia
    c. Paraplegia from trauma injury 10 years prior
    d. Renal failure

27. Patients with pseudocholinesterase deficiency will;
    a. demonstrate acute sensitivity to non-depolarizing neuromuscular blocking agents
    b. demonstrate acute insensitivity to depolarizing neuromuscular blocking agents
    c. demonstrate prolonged duration of action with non-depolarizing neuromuscular blocking agents
    d. demonstrate prolonged duration of action with depolarizing neuromuscular blocking agents

28. A succinylcholine induced phase II block will act very similar to;
    a. non-depolarizing neuromuscular blockade
    b. shortened depolarizing neuromuscular blockade
    c. approximately twice the typical SUX duration of action
    d. approximately twice the duration of a depolarizing neuromuscular blockade

29. Malignant hyperthermia is frequently associated with succinylcholine administration. The pathology rests with a disorder involving the ryanodine receptors on the endoplasmic reticulum causing a large calcium release. Your best therapy would include;
    a. calcium channel blockers, rapid cooling, maximized oxygenation and ventilation
    b. maximized oxygenation and ventilation, rapid cooling and dantrolene sodium administration
    c. rapid cooling, dantrolene sodium administration and potassium supplementation
    d. maximized oxygenation and ventilation, rapid cooling and administration of a non-depolarizing neuromuscular blocker.

30. The earliest clinical signs to identify malignant hyperthermia are;
    a. profound temperature spike unresponsive to cooling methods
    b. obvious myoclonus and trismus after SUX administration
    c. ventricular arrhythmias and peaked T-waves on the ECG
    d. profoundly rising end-tidal CO_2 levels unresponsive to increasing Ve

31. Common complications of surgical cricothyroidotomy include all but which of the following;
    a. bleeding
    b. false passage to the mediastinum
    c. false passage to the esophagus
    d. aortic injury

32. Which of the following is not a non-depolarizing neuromuscular blocking agent?
    a. succinylcholine (*Anectine*®)
    b. vecuronium (*Norcuron*®)
    c. rocuronium (*Zemuron*®)
    d. cisatracurium (*Nimbex*®)

33. Depolarizing agents;
    a. work by competing with acetylcholine (ACh) for the ACh receptor sites without any intrinsic action on those sites
    b. work by competing with acetylcholine (ACh) for ACh pre-synaptic sites stopping the release of ACh from those sites
    c. work by competing with acetylcholine (ACh) at the ACh receptor and agonize those receptor sites
    d. work by agonizing the acetylcholine (ACh) receptors like ACh but in a non-competitive manner

34. When non-depolarizing neuromuscular blocking agents are used in high dosages to facilitate RSI conditions, you should expect;
    a. shorter duration paralysis as liver enzymes are induced (i.e.-CP450)
    b. faster onset of muscle relaxation with longer duration
    c. faster onset of sedation with muscle relaxation
    d. better intubation conditions versus the conditions seen in patients with depolarizing neuromuscular blockade use

35. The "supra-glottic" class of airways include all but which of the following?
    a. laryngeal mask airway (LMA)
    b. Esophageal-tracheal CombiTube®
    c. King LT®
    d. PAXpress® (PAX)
    e. Cuffed Orotracheal Airway (COPA®)
    f. Cobra Perilaryngeal Airway
    g. Gum Elastic Bougie (GEB)

36. Supra-glottic airways are typically __________ and __________.
    a. inserted 'blindly' , not truly patent airways
    b. inserted under direct visualization , inflated to their designated pressure
    c. inserted 'blindly' , require continuous training to remain proficient
    d. inserted 'blindly' , commonly fail to provide adequate oxygenation

37. You have just performed an RSI technique utilizing lidocaine, atropine, fentanyl, etomidate and succinylcholine to successfully intubate a 4-year-old male presenting in acute status asthmaticus. His pre-induction SaO_2 was 93% on a non-rebreather mask at 15lpm flow. While you are taping the tube your partner says, "Oh Sh__!" You look up and note the cardiac monitor to show VF. You should;
    a. identify the arrest as hypoxic in etiology, aggressively ventilate and oxygenate, follow PALS recommendations for the resuscitation
    b. identify the arrest as hyperkalemic in etiology, aggressively ventilate and oxygenate, administer CaCl, $NaHCO_3$, insulin and D50 immediately as you begin the resuscitation
    c. identify the arrest as a polypharmaceutical reaction, aggressively ventilate and oxygenate, administer naloxone (*Narcan*®), flumazenil (*Romazicon*®) and follow PALS recommendations for the resuscitation. Consider an allergic reaction as a high probability.
    d. identify the arrest as unknown in etiology, aggressively ventilate and oxygenate while proceeding along PALS recommendations to rule out all causes in an orderly fashion.

38. An absolute contraindication to blind nasal intubation is;
    a. maxillo-facial trauma
    b. basilar skull fracture
    c. cervical spine injury
    d. apnea

39. The proper depth of a #7.0mm endotracheal tube would be estimated at;
    a. 19cm
    b. 21cm
    c. 23cm
    d. 25cm

40. If a defasciculating dose of a non-depolarizer is administered while preparing to RSI, you should anticipate;
    a. using an increased dosing of SUX at the time of induction
    b. significantly increased duration of paralysis post SUX administration
    c. accelerated sedation from the induction agent(s)
    d. increased positive pressure ventilation prior to intubation

KEY & RATIONALE

AIRWAY MANAGEMENT & PHARMACOLOGY

1. A Obviously this is a stupid question. I placed this here to make a very important point. As we (transport providers) become more educated in medicine we find that the 'black and white' fades to shades of grey. You will undoubtedly find yourself considering other therapies prior to airway management in just the right scenario. When taking these certification exams, they WILL NOT account for those same variables so when presented with two choices and airway is one, ALWAYS, ALWAYS, ALWAYS choose airway first.

TEST TIP
AIRWAY is ALWAYS FIRST!

2. D Intubation is "indicated" in all of these patients. While you may opt not to intubate some, it is nonetheless indicated or appropriate should you choose to go that route. Consider intubating anyone that has; a current airway patency issue (i.e.- the heroin user), a failure to oxygenate or ventilate (i.e.- the AMI victim is not oxygenating while he is ventilating. An asthmatic may have a ventilation failure while he oxygenates just fine), finally consider intubating based on expected clinical course (i.e.- The house fire victim is currently ventilating well and likely oxygenating acceptably, however his injuries and circumstance suggest airway closure from edema is imminent and pre-emptive intubation is recommended)

3. A The algorithms utilized for the exam come from the Air & Surface Patient Transport, Principles & Practice text by Renee Holleran. She obtained those algorithms from the Manual of Emergency Airway Management by Ron Walls. You should make yourself familiar with these algorithms prior to testing.

4. D One of the primary principles and benefits of the emergency airway algorithms is that all algorithms lead to the failed airway algorithm of the airway is not successfully managed using the initial algorithm. You never 'jump' from crash to difficult or RSI to crash, etc.

5. C The first two criteria are used in all algorithms to dictate when the Failed airway algorithm should be initiated. Initially finding a patient unconscious and unresponsive near death dictates the Crash Airway algorithm should be initiated. As a rule, the Failed Airway is never the first algorithm on initial airway management unless you are incorporating other provider's airway management into the decision process. I.e.- you arrive and a qualified airway manager has already attempted intubation three or more times without success, you could reasonably and arguably move to the Failed Airway as "your" first algorithm for management.

TEST TIP

Can't keep SpO_2>90% <u>OR</u> 3 failed intubation attempts?

⇓

Use FAILED AIRWAY algorithm

6. D If the patient does not require sedation, RSI is not indicated. If you cannot reliably ventilate the patient with a BVM, RSI is not indicated. If you cannot expect to successfully intubate them, RSI is not indicated. If your patient doesn't meet all three of these criteria, RSI is hazardous, not warranted and not legally defensible. The Crash Airway allows for paralyzation without sedation as sedation would not be indicated in someone unconscious/unresponsive, near death. RSI will improve intubation success rates in patients appropriately selected for RSI. RSI statistically will drop airway management success rates for patients in which RSI was inappropriately selected.

7. A All of the choices given are good for assessing BVM compliance in most patients. However, accurate SpO_2 is the "gold standard" for assessing success of oxygenation using the BVM. Notice I said "oxygenation" not "ventilation". The "gold standard" in assessing ventilation is capnography.

8. C The Cormack-Lahene Grading system is utilized upon invasive visualization of the glottic opening. The MMT is performed using an external exam, see diagram next page. The ULBT is another external exam with increased sensitivity over the MMT. The 3-3-2-1 exam is an external exam assessing for; 3 finger width opening of the mouth, three finger width thyromental distance, 2 finger width thyrohyoid height and one finger width forced under-bite at the teeth.

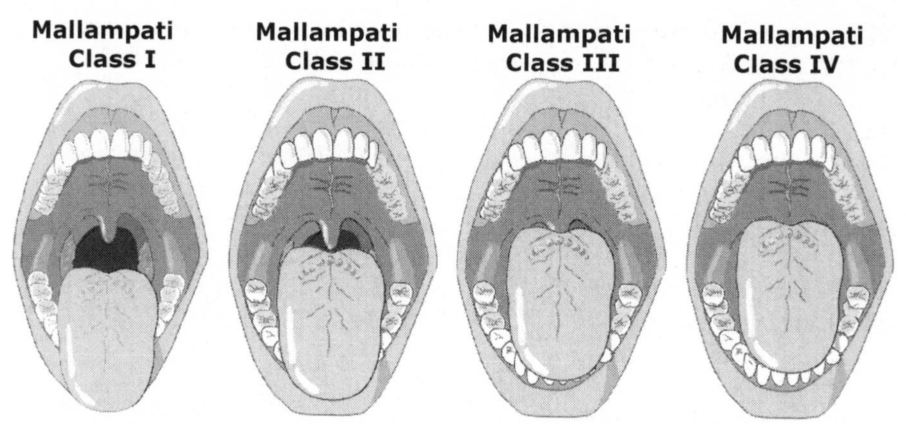

Class I – able to visualize the palatoglossal and palatopharyngeal arches, the entire uvula, superior and inferior tonsilar pillars

Class II – able to visualize the palatoglossal and palatopharyngeal arches, the entire uvula, with loss of the majority of the tonsilar pillars

Class III – visualization of the palatoglossal and palatopharyngeal arches is lost, the tip of the uvula cannot be seen and the tonsilar pillars are not seen

Class IV – only visualization of the hard palate and tongue is possible

9. B Protruding lower incisors are not typically troublesome whereas protruding upper incisors are. The latter because of their likelihood of being used as a fulcrum in levering of the laryngoscope during instrumentation as well as oropharyngeal access limitation. All other criteria there should prompt you to consider the airway potentially difficult.

10. B This patient should prompt you to manage them via the Difficult Airway algorithm. From the information provided you can use the Difficult Airway algorithms option for RSI with a "double-setup". Most anesthesia literature and texts will recommend a Miller blade for the difficult airway but the practitioner should make their initial attempt their best and thus use the blade, which they most often find success. Option A. is too cavalier an approach; option C. too conservative for an "emergent intubation" but requesting fiber-optic equipment/personnel be summoned would not be a poor plan. D. is a very attractive option however following the difficult airway algorithm, you would not be directed to this technique initially. Also the Macintosh blade is not definitively going to provide "optimal visualization", in fact it is a more technical blade to use and frequently more difficult to use, as designed, for the new airway manager.

11. A Sellick's maneuver is intended to obstruct the esophagus thus preventing passive regurgitation and subsequent aspiration. In some patients, Sellick's maneuver (cricoid pressure) does provide a better view. However studies have shown that Sellick's maneuver will in fact create a worse view in many patients and thus should not be routinely utilized. BURP maneuver or 'backward upward rightward pressure' (from the patients perspective) is intended to bring the larynx into truer axis alignment and improve glottic visualization as will lifting the head to a true sniffing position (see diagram below).

12. B RSI was created for the rapid intubation of the patient with a full stomach. The entire premise of RSI is to prevent active vomiting with subsequent aspiration. Studies have demonstrated that RSI of the poor RSI candidate actually causes a decrease in successful intubation rates. While appropriate RSI will create good intubation conditions as well as prevent movement and provide amnesia, those effects are not "why" we use RSI however.

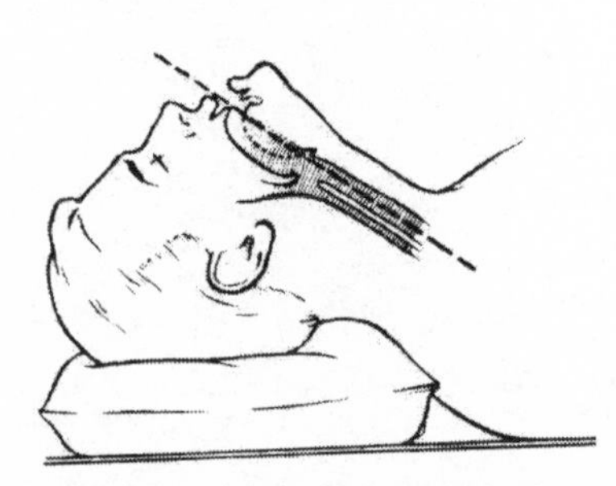

13. D Etomidate (*Amidate*®) is the best drug to choose for an awake technique from the options provided. One must consider the following points: onset time, duration, emergence time, respiratory depression, hemodynamic stability, and side effect profile. Ideally we would like an agent with rapid onset, predictable short duration, rapid emergence without respiratory or hemodynamic depression and with no side effects at all. Midazolam cannot be used reliably without significant hemodynamic complications and respiratory depression. Diazepam carries the same complications as midazolam with a longer more unpredictable onset and duration. Ketamine has a relatively short onset and duration and while respiratory drive sparing and hemodynamically stimulating in most cases, it's emergence is fraught with complication, as well as the side effect profile is less attractive considering the hypersalivation issues. Etomidate provides a rapid onset, predictable, short duration and rapid emergence. Respiratory depression is rare and hemodynamic stability is unsurpassed by any other induction agent currently available. The side effect profile is minimal with short and long term issues being insignificant comparatively to the other agents discussed.

14. D "Secondary" confirmation methods are methods which employ a mechanical device to provide an "objective" assessment. Option A is the best "primary" confirmation being that it is a "subjective" assessment. The best mechanical or "objective" confirmation is capnography or waveform display of end-tidal CO_2 elimination.

15. A SaO_2 and PaO_2 should have "optimized" long before you would have replaced all the nitrogen (denitrogenated) with oxygen in the FRC. While C may occur with better oxygenation it is not the purpose of such. Replacing all the nitrogen in the FRC with oxygen allows the patient to maintain oxygen saturation during apnea for longer periods of time.

16. B Contrary to the Paramedic texts a NRBM at high flow will not provide the purported 80-100% FiO_2. In fact research demonstrates we commonly only receive 40-60% with that technique due to poor mask seal/fit and device function. Optimal pre-oxygenation would be obtained allowing the patient to breath via a BVM with a tight mask seal (assuming a recently designed BVM using a passive or 'duck bill' valve system between the bag and the exhalation port of the device.) Device dead space and flow restriction is minimal. Optimally we should not 'assist' the patient's breathing to prevent any gastric insufflation prior to intubation.

17. C Five minutes is the suggested pre-oxygenation minimum. It has also been suggested that five to eight vital capacity breaths in rapid succession accomplishes the same task.

18. D Failed airway algorithm is the only algorithm with supra-glottic rescue devices recommended as the sole airway option. Intubating laryngeal mask airways (I-LMA's) are permitted in the difficult airway algorithm assuming the patient is intubated via that device.

19. A When considering who will desaturate the fastest, consider the following factors in order of slowest to desaturate to fastest. Illness ⇒ Pediatrics ⇒ Obesity. The obese desaturate much faster than patients with a normal or thin habitus. Pediatric patients desaturate faster than adults generally speaking and finally the sick will desaturate faster than the healthy. So the worst-case scenario is an obese pediatric patient that is very sick. Of the choices you should note that option A is a 3 foot, 8 inch tall 8 year old that weighs almost 150 pounds. That's one chubby little kid!

All the other candidates are fairly height-weight proportionate and disease states are relatively similar in terms of oxygen demand.

20. B The most common agents to precipitate an anaphylactic reaction are the NMBA's, especially succinylcholine. Occurrence rates are very high with antibiotics (especially penicillin) and latex but NMBA's are the most common. You should note that many training scenarios, especially the ACLS of previous years talked repeatedly about lidocaine allergies. In fact an allergy to lidocaine is almost unheard of with only a few documented cases in the literature ever confirmed. Lidocaine (*Xylocaine®*) is an amide, allergic reactions to local anesthetics come from the ester based solutions such as procaine (*Novacaine®*). Assuming all the 'caines' are the same is a mistake, the two classes (esters and amides) are distinctly different.

21. D The induction agent and muscle relaxants should be selected and administered in such a manner that they alone provide adequate sedation, amnesia and muscle relaxation. While the opioid is additive to the sedation it is much less useful in regard to amnesia and muscle relaxation. The primary reason for opioid use prior to intubation is to alleviate the pain response ("sympathetic dump") to instrumentation of the airway itself.

22. B Succinylcholine is noted to precipitate bradycardia, especially in pediatric patients and repeat administrations, making A incorrect. Ketamine will typically prompt a sympathetic response or augmentation of such thus causing tachycardia and hypertension, therefore C is a poor choice. While instrumentation of the airway does commonly cause tachycardia, prevention of such is performed with opioids not atropine. In fact atropine would make this worse. Ketamine is known to cause hypersalivation via a parasympathetic efferent pathway. Utilizing atropine will antagonize or block this pathway thus preventing the salivation. As a side note, utilizing atropine for this will also cause the tachycardia precipitated by the ketamine to possibly worsen (increasing myocardial oxygen demands). A better agent to use in this case as an antisialogogue would be glycopyrrolate (*Robinul®*).

23. D Defasciculating doses of a non-depolarizing medication will attenuate the ICP spikes associated with laryngoscopy according to most studies. Myalgias were once believed to be attenuated by defasciculating pre-medication but research does not support this theory, in fact myalgias will undoubtedly occur in young males, especially those with lean, muscular habitus regardless. Potassium levels will elevate regardless of defasciculation thus you should never give succinylcholine (SUX) to anyone with a known elevated potassium level. You can anticipate a standard intubation dose of SUX will elevate their potassium by ~0.5mEq/L. Malignant hyperthermia can only be prevented by complete avoidance of MH triggering agents.

24. D Fentanyl commonly causes coughing and sudden respiratory arrest when given as described. "Rigid-chest syndrome" has been well documented with fentanyl but is more commonly described with fentanyl's more potent relatives, sufentanil and remifentanil. Antegrade amnesia is not *reliable* with any of the opioids and retrograde amnesia is not reliably produced by anything short of a head injury.

25. D The literature has documented the occurrence of adrenal insufficiency with the administration of a single intubating dose of etomidate (*Amidate*®). While this decrease in cortisol production does not appear to be significant especially in light of the benefits of etomidate use prehospital, it is of concern in some specific patient populations. Etomidate is no longer used as a continuous drip for sedation due to this side effect. Etomidate also causes an increased occurrence of nausea and vomiting after emergence. Myoclonus is occasionally seen but not considered prohibitory in it's use, especially in the RSI scenario. Etomidate's respiratory and hemodynamic profiles make it the most attractive prehospital induction agent.

26. D SUX is contraindicated in all muscular disorders such as Duchenne's, ALS, multiple sclerosis, Guillian-Barre, Charcot-Marie-Tooth, etc. as well as any known pre-existing hyperkalemia. SUX may be used in renal failure patients that have a normal potassium level at the time of administration. Do not assume renal failure equals hyperkalemia. Any person with a disease/injury that would result in denervation should not have SUX administered to them, i.e.- significant burns, crush injuries, major trauma, cord injuries, CVA's, etc. Even patients that are simply immobile in an ICU bed for as little as 24-48 hours may be predisposed to upregulation of immature ACh receptors and thus susceptible to lethal hyperkalemia with SUX administration. Any history of malignant hyperthermia in their family should contraindicate the use of SUX.

27. D Pseudocholinesterase (aka- plasmacholinesterase, butyrylcholinesterase) is the component in plasma which biotransforms succinylcholine to an inactive state. Deficiency of this enzyme will cause a standard SUX dosing to have a duration of effect of two to eight hours. This disorder occurs in about one in 25 North Americans, interestingly it occurs in about seven out of ten middle-easterners.

28. A The phase II block phenomenon associated with SUX administration is not fully understood. The putative mechanism is thought to be an actual blocking of the ion channel gate closure by the SUX molecule itself causing an extensively prolonged neuromuscular blockade. Evaluation with a peripheral nerve stimulator will find very similar results (TOF fade and tetanic fade) to those seen with non-depolarizing blockade as well as very similar duration of action.

29. B First recognize that dantrolene sodium (*Dantrium*®) is the drug of choice for reversing malignant hyperthermia (MH). If you only remember that you can narrow down the choices to B and C. Also try to remember that these patients will enter a very high metabolic rate very quickly (this is what causes the hyperthermia). This causes a massive increase in oxygen needs and CO_2 production, thus matched oxygenation and ventilation is critical, commonly requiring as much as four times their normal minute volume. Finally recall that massive stimulus of muscles will cause massive shifts of potassium creating a hyperkalemic state. You would want to treat this with the standard protocol ($NaHCO_3$, insulin, dextrose, albuterol, etc.) less the calcium and then provide aggressive fluid resuscitation. Finally maximizing renal output is important so along with aggressive fluid resuscitation, frequently the use of furosemide and/or mannitol is employed to manage myoglobinuria secondary to rhabdomylosis.

30. D Refer to the MH explanation above. The first signs are elevating $ETCO_2$ and dropping SpO_2 that responds poorly to elevations in Ve or FiO_2. These will commonly, but not always, be accompanied by myoclonus and/or trismus. However SUX administration commonly precipitates short periods of myoclonus and trismus *without* any MH occurrence so these signs alone are not definitive. Temperature elevation is a late sign of MH. Arrhythmias and peaked T-waves indicative of hyperkalemia are commonly seen after the oxygenation-ventilation issues.

31. D Aortic injury is not a common complication of surgical cric's due to it's inferior location from the cricothyroid membrane. If you hit the aorta doing a cric, you're too low... way too low. All other complications listed here are relatively common along with passage of the ETT into the oropharynx, vocal cord injury, thyroid injury and epiglottic injury.

32. A Succinylcholine is the only depolarizing neuromuscular blocking agent currently available in the USA. All the other agents listed are non-depolarizing agents from one of the two sub-families of such. All of the "-ronium's" are aminosteroid agents while all of the "-curium's" are benzylisoquinoliniums.

33. D Succinylcholine, a depolarizing paralytic is also commonly referred to as a 'non-competitive neuromuscular blocking agent' because it's mechanism of action is to mimic endogenous ACh. All of the non-depolarizing paralytics are called 'competitive neuromuscular blocking agents'. The non-depolarizers physically compete with ACh for the ACh receptor, but do not possess the intrinsic action within them to trigger the receptor, they simply block it. Test tip, if you were 'shooting blind' on this question and had absolutely no idea you might consider the following. Looking at each of the four answers you see there is only one that offers the option of "non-competitive". If you're *shooting blind* try to identify patterns including "which one's not like the others".

34. B The whole idea behind using high dosages of non-depolarizers is to achieve rapid onset similar to SUX. The trade off is that in that high a dosage, duration will be prolonged over standard dosing of the same agent. D may be an attractive option but in fact this is not true for *all* patients. Opting to administer high doses of paralytics to poor candidates can actually create worse intubation conditions and poorer outcomes.

35. G All of the devices listed except the bougie are supra-glottic devices. I included this question to be sure you are at least familiar with the names of the most common rescue devices available. The bougie is actually an intubation assisting device and not an airway of any kind. I would recommend you review a basic picture of each of these devices if you are not familiar with them prior to testing. I would suggest you carry a bougie with you anytime you're on duty.

36. A All of the supra-glottic airways are intentionally designed to be inserted without direct visualization. After all, if direct visualization were possible, why would you use them? You'd place an endotracheal tube instead..... right? While most of the supra-glottic airways require inflation after insertion, they do not require high levels of training to maintain acceptable proficiencies and have been shown to provide acceptable levels of oxygenation and ventilation. While these rescue airways have obvious benefits, none provide a truly 'patent airway' which is defined as a tube in the trachea with an inflated cuff that continues to hold pressure.

37. B A is tempting, especially considering the original complaint and can't be ignored. C is a distinct possibility however *Narcan*®, flumazenil and an allergic reaction are already treated or negated by the fact your patient is already intubated and about to receive a whopping dose of epinephrine for the arrest IV. This makes C a very unlikely cause. D is a good all around generic approach and quite acceptable for the standard transport provider. The key points here are SUX, toddler, male, sudden arrest. This is the typical scenario surrounding sub-clinical Duchenne's muscular dystrophy. It is this exact scenario that prompted the FDA to place a black box warning on SUX and anesthesia's recommendation that "succinylcholine should not be used for *routine* intubation in pediatrics". Duchenne's typically presents in males and is commonly fatal by eight years of age. Parents may report that the child "walks like a duck" or other peculiar abnormalities of gait. You may also notice very prominent calf muscle development. These patients have a proliferation of immature ACh receptors and administration of SUX can cause an overwhelming release of potassium extracellularly.

38. D The only true absolute contraindication is apnea. The ability to hear ventilation via the ETT during placement is the only safe way to perform blind nasotracheal intubation (BNTI). Maxillo-facial trauma and basilar skull fracture do increase the hazards of BNTI and should be considered strong relative contraindications however the literature has demonstrated that penetration of the cranial vault is rare and has only been documented in instances of exceedingly poor judgment and technique. For many years, cervical spine injuries were thought safest intubated using the BNTI technique. It was subsequently proven that cautious in-line intubation could achieve endotracheal intubation with the same safety margins. Today, awake BNTI is still typically used in the operating room as the preferred intubation technique for unstable cervical spine surgeries.

39. B Endotracheal tube depth can be estimated by multiplying the diameter times three. (i.e.- 7 X 3 = 21) This works for adult and pediatric ETT's. Again this is an 'estimate'.

40. A When you administer the non-depolarizing NMBA pre-medication you block *some* of the ACh receptor sites. For SUX to achieve flaccid paralysis it must trigger an overwhelming majority of the body's sites to precipitate the phenomenon of muscle exhaustion that creates the flaccidity. To make it even simpler consider that non-depolarizers and depolarizers work in direct opposition. SUX dosing should be slightly higher after defasciculating doses are used. Duration of SUX is not significantly altered and sedation is in no way affected by any of the NMBA's. You would not want to positive pressure ventilate your RSI candidate at all prior to intubation. The procedure is: prepare ⇒ pre-oxygenate ⇒ premedicate ⇒ induce/relax ⇒ intubate....then, and only then... ventilate.

TRAUMATIC INJURY & HEMATOLOGY

1. A late sign of tension pneumothorax would be;
   a. low SpO_2
   b. tachycardia
   c. jugular venous distention
   d. tracheal shift

2. The mechanism by which tension pneumothorax causes hemodynamic collapse is;
   a. gross/overwhelming hypoxia
   b. compression of the myocardium
   c. compression of the great veins
   d. collapse of the unaffected lung

3. Signs of massive hemothorax include all but which of the following?
   a. tachycardia
   b. tracheal shift
   c. hypotension
   d. altered level of consciousness

4. For an open chest wall injury (sucking chest wound) to preferentially entrain air (to "suck"), it must be at least _____ the size of the narrowest part of the airway.
   a. 1/2
   b. 1/4
   c. 3/4
   d. 1/3

5. The best initial therapy for an open chest wall defect in a spontaneously breathing patient is;
   a. high flow oxygen administration
   b. occlusion of the defect with an occlusive dressing at end-inspiration
   c. occlusion of the defect with an occlusive dressing at end-expiration
   d. needle decompression

6. Needle decompression may be performed at all but which of the following;
   a. 2nd intercostal, mid-clavicular line
   b. 4th intercostal, anterior axillary line
   c. 5th intercostal, mid-scapular line
   d. 5th intercostal, mid-clavicular line

7. The best indication for decompression of a tension pneumothorax is;
   a. radiographic confirmation of pneumothorax
   b. tracheal shift
   c. difficulty breathing
   d. absent lung sounds on one side

8. Thoracic tube thoracostomy should be performed;
   a. 2nd intercostal space, mid-clavicular line
   b. 4th intercostal space, anterior axillary line
   c. 5th intercostal space, mid-axillary line
   d. 5th intercostal space, anterior axillary line

9. Emergent treatment of the patient with a tension pneumothorax will include all of the following choices in many situations. Which one of the choices would be best to avoid if at all possible.
   a. tube thoracostomy
   b. needle decompression
   c. intubation
   d. positive pressure ventilation

10. Which of the following is true concerning thoracic injuries?
    a. only penetrating trauma above the level of the diaphragm should be considered "thoracic trauma"
    b. tension hemothorax will present with tracheal shift in the late stages
    c. RSI and intubation should be considered early in all thoracic trauma victims
    d. cyanosis is commonly a late sign in thoracic trauma

11. Which of the following is false concerning thoracic trauma?
    a. tracheal shift at the carina is an early sign of pneumothorax on radiographic exam
    b. pericardial tamponade typically presents with tachycardia in the early stages
    c. thoracic trauma should always include a high index of suspicion for concomitant hypovolemia
    d. when performing a tube thoracostomy, the tube should be placed as close to the inferior margin of the costal as possible

12. While examining a patient for transfer to a trauma center you evaluate the patient's chest film. The AP chest film demonstrates that a chest tube has been placed in the left pleural space. The radiopaque indicator's 'break' is at the most lateral edge of the pleural space and you note significant air in the subcutaneous tissue at the point of entry as well as a persistent, however small, pneumothorax. This would most likely indicate;
    a. The chest tube is properly placed and functioning appropriately
    b. The chest tube is properly placed and functioning improperly
    c. The chest tube is properly placed but likely occluded or not to suction
    d. The chest tube is not properly placed and functioning improperly

13. You have been called to the scene of a school shooting in which your patient was the victim of a close range shotgun wound to the right chest. On arrival you find the patient intubated by the SRT medic, being ventilated with 100% oxygen. Another SRT member is holding their hand over the chest wound stating, "when I take my hand off it's making a sucking sound!" Which of the following describes the proper management of this patient's open chest wound.
    a. occlusive dressing placed on end-exhalation, taped on all sides
    b. occlusive dressing placed on end-exhalation, taped on three sides
    c. occlusive dressing placed on end-inhalation, taped on all sides
    d. occlusive dressing placed on end-inhalation, taped on three sides

14. Later in the management of the patient described in the prior question you have placed a chest tube on the right side with positive air and blood return. After attaching the tube to an appropriate negative pressure collection device you would need to;
    a. needle decompress at the 2nd intercostal to complete the vent aspiration circuit
    b. finish taping the occlusive dressing at the initial wound site (tape the fourth side)
    c. aggressively hyperventilate the patient
    d. turn the patient injured side up to assist venting of the pneumothorax

15. You have identified a possible flail segment on your motor vehicle crash victim. Which of the following would not be beneficial in treating the flail segment?
    a. fluid restriction
    b. positive end expiratory pressure
    c. sand bag splinting of the segment
    d. turn the patient injured side down

16. Early signs of pericardial tamponade include;
    a. Trousseau's sign
    b. pulsus paradoxus
    c. Jerisch-Bergen reflex
    d. widening pulse pressure

17. The primary treatment for early pericardial tamponade is;
    a. preload augmentation
    b. afterload reduction
    c. augment contractility
    d. pericardiocentesis

18. Which of the following is not part of Beck's Triad?
    a. muffled heart tones
    b. hypotension
    c. narrowing pulse pressure
    d. jugular venous distension

19. A generalized harsh systolic murmur would be indicative of which pathology?
    a. acute left ventricular myocardial infarction
    b. pericardial tamponade
    c. pulmonary embolism
    d. aortic rupture

20. A scaphoid abdomen would be indicative of which pathology?
    a. liver laceration
    b. diaphragmatic rupture
    c. diverticulitis
    d. bowel intussusception

21. You are transferring a patient with a diagnosis of 'caval laceration, rule out liver laceration' secondary to a knife assault. The patient has been fluid resuscitated, blood administered and stabilized nicely. Current vitals are HR 92, BP 104/56, RR 18, SpO_2 98% on an FiO_2 of 0.5. You currently have an 18g angiocath at the right antecubital vein, a 16ga angiocath at the left antecubital vein and a dual-lumen CVP cath at the right subclavian. During transport you would consider;
    a. maintaining systolic BP greater than 110mm Hg, use pressors PRN
    b. obtaining IV access below the level of the diaphragm
    c. attaching a blood set with pressure transfuser to one of the lumens on the central line
    d. intubating the patient

22. The primary finding indicative of aortic disruption on an AP chest X-ray is;
    a. "patchy infiltrates"
    b. "flattened diaphragm"
    c. "Kerley's B lines"
    d. "widened mediastinum"

23. Which of the following is most likely to be said by a patient with an aortic aneurysm?
    a. "My pain is dull, located under my sternum and radiates to my arms and jaw. It's constant and hurts terribly!"
    b. "My back is killing me. It feels like it's ripping apart between my shoulder blades. Can't you help?!"
    c. "My pain gets worse when I take a deep breath. It's sharp and seems to come and go. If I can just sit up that will help."
    d. "Anyurism? What's my butt got to do with this? It's my chest that hurts!"

24. Which of the following therapies is a high priority for your diaphragmatic rupture patient.
    a. NG/OG tube placement and evacuation
    b. foley placement and output monitoring
    c. oxygen via nasal cannula at 4-6lpm
    d. large-bore IV access with aggressive fluid resuscitation

25. Which of the following is not a classic sign of diaphragmatic rupture?
    a. scaphoid abdomen
    b. bowel sounds in the thoracic cavity
    c. dyspnea
    d. diarrhea

26. Your patient is a reportedly intoxicated college student involved in an assault at a county recreational park. Per bystanders, he was reportedly running across the park when another male struck him in the chest with a croquet mallet and 'dropped him'. You have now placed a chest tube on the affected side of your tension pneumothorax victim, however you note a continuous air leak even after verifying tube depth, system integrity and seal around the tube insertion itself. The most likely cause for this would be;
    a. esophageal perforation
    b. diaphragmatic air
    c. torn endotracheal tube cuff
    d. tracheobronchial disruption

27. Therapy to resolve the situation described in the question above would be;
    a. NG/OG tube insertion
    b. position the patient reverse trendelenburg as he will tolerate
    c. replace the endotracheal tube
    d. perform a mainstem intubation and adjust tidal volume

28. Your patient, Stanley the Magnificent, has been diagnosed with a proximal esophageal perforation after attempting to break his record of swallowing twelve swords at the carnival. Which of the following therapies would not be indicated?
    a. NG/OG tube insertion
    b. antiemetic administration
    c. antibiotic administration
    d. aggressive use of morphine for pain as hemodynamics will tolerate

29. In physics, the law which states, "For every action there is an equal and opposite reaction" would be;
    a. Newton's first law of motion
    b. Newton's second law of motion
    c. Newton's third law of motion
    d. Einstein's theory of relativity

30. Your patient was an unrestrained driver in a head-on collision. As you approach the wreckage the first responder says, "Looks like he did an 'up and over'." referring to the mechanism of injury. Based on this, which body organ injury pattern is most likely to present.
    a. spleen, liver, bowel, pelvis
    b. brain, spine, lungs, heart
    c. femurs, pelvis, bowel, liver
    d. face, upper arms, brain, spine

31. Your patient was ejected from a motorcycle at high speed without a helmet. She has been diagnosed with LeFort I, II and III fractures bilaterally. The most common presenting sign for this injury pattern is;
    a. epistaxis
    b. otorrhea
    c. periorbital ecchymosis
    d. trismus

32. For the patient described in the prior question, which of the following management strategies is contraindicated?
    a. intubation
    b. NG tube placement
    c. c-spine stabilization
    d. foley placement

33. Your patient has presented with massive facial and head injuries. On initial contact you note the face feels "loose" and upon attempting to bag the patient you note central features (the nasal area) seem to be easily compressible and "floating". The patient is unresponsive and exhibiting trismus. The decision has been made to place an endotracheal tube. Which of the following meds is most likely to permit successful and appropriate intubation.
    a. benzodiazepine (i.e.- midazolam)
    b. opioid (i.e.- fentanyl)
    c. neuromuscular blocker (i.e.- succinylcholine)
    d. none of these will reliably permit successful intubation

34. Your patient was the restrained driver of a car struck from behind. Which of the following injury patterns does not coincide with the mechanism of injury?
    a. T12-L1 spinal injury
    b. C2 spinal injury
    c. ankle fracture
    d. wrist fracture

35. The two classic injury patterns described for motorcycle crash victims are;
    a. down and under ; rider jump
    b. up and over ; lay it down
    c. step off ; endo
    d. 270 quarter spin with a twist ; cannon ball

36. Which of the following accident categories results in the most deaths each year?
    a. head on collision
    b. side impact collision
    c. rollovers
    d. rear-end collisions

37. Management of the trauma victim commonly follows the standards of "ABC's", Airway, Breathing, Circulation. However we must also manage cervical spine injuries early as well. At which step is C-spine typically managed?
    a. with Airway
    b. with Breathing
    c. with Circulation
    d. after the ABC's

38. Blood pressure will not typically show a drop until the trauma victim experiences a blood loss of more than;
    a. 10-20%
    b. 20-30%
    c. 30-40%
    d. 40-50%

39. When performing a fluid resuscitation, the best IV catheter to use for fastest fluid infusion would be;
    a. 18ga 1.25" peripheral IV
    b. 16ga 1.16" peripheral IV
    c. 18ga 18cm central venous catheter
    d. 20ga 2.25" arterial catheter

40. When performing a fluid resuscitation, the best IV solution to use in the acute phase would be;
    a. D5NS
    b. Albumin
    c. Ringer's Lactate
    d. 3% Hypertonic Saline

41. The best/most reliable clinical sign to indicate onset/progression of shock in your trauma patient would be;
    a. increase in heart rate
    b. decrease in blood pressure
    c. decrease in oxygen saturation
    d. drop in level of consciousness

42. Successful trauma management is built upon managing of the ABC's, aggressive fluid resuscitation, rapid transport to definitive care and ________.
    a. blood administration
    b. constant reassessment
    c. rapid immobilization
    d. ventilatory support

43. Common problems associated with farming accidents include all but which of the following;
    a. hazardous material spill
    b. heavy equipment with lower center of gravity
    c. equipment operator entrapment
    d. prolonged delays in arrival of medical care

44. Typical fall related injuries in adults commonly have a series of injuries referred to as the "lover's leap" syndrome. This includes all but which of the following;
    a. cervical and thoracic axial loading injuries
    b. Colles' fractures
    c. calcaneus fractures
    d. deceleration injuries to internal organs

45. Your patient was stabbed by his estranged wife. You would anticipate that the trajectory of the stab wound was;
    a. random
    b. up and to the left from the initial penetration
    c. down from the initial penetration
    d. up and to the right from the initial penetration

46. High velocity bullets are classified as any projectile traveling;
    a. faster than 1000 feet per second (fps)
    b. faster than 2000 feet per second (fps)
    c. faster than 3000 feet per second (fps)
    d. faster than 5000 feet per second (fps)

47. Ideally, all of the information pertaining to a gunshot wound is pertinent for patient care except;
    a. type of firearm involved
    b. bullet construction
    c. distance from firearm when struck
    d. manufacturer of bullet/projectile

48. Your patient was injured by an improvised explosive device (home-made bomb). During the explosion the patient received a coup-contrecoup injury to the head upon striking the ground. This injury would be considered a;
    a. primary injury
    b. secondary injury
    c. tertiary injury
    d. quarternary injury

49. Your 80kg patient has been involved in a house fire due to a methamphetamine lab explosion. He has received first-degree burns to the abdomen and his lower back. He has second-degree burns to the remainder of the chest and upper back. He has third degree burns of the face and head, hands, and arms. You would calculate his BSA burned as;
    a. 45%
    b. 63%
    c. 39%
    d. 32%

50. While caring for the patient described in the previous question you would calculate his total fluid resuscitation volume per the Parkland Burn Formula to be;
    a. 12.4 liters
    b. 24.4 liters
    c. 4.4 liters
    d. 14.4 liters

51. Continuing our care of the patient in the previous two questions, you would want to anticipate all but which of the following;
    a. cyanide poisoning
    b. carbon monoxide poisoning
    c. methemoglobinemia
    d. psychiatric disturbance

52. Assuming the patient above is producing urine and demonstrates he is not 'shocky', you would set your initial IV fluid administration rate at;
    a. 500ml/hr
    b. TKO
    c. 250ml/hr
    d. 900ml/hr

53. You delivered the burn patient referenced in the above series of questions to your local hospital. The receiving staff has now requested you return and transfer the patient to a tertiary burn center. It has been seven hours since the initial scene call. Upon arrival for the transfer you note during their report, that since the patient was producing urine on initial arrival, the IV's were set to 120ml/hr

maintenance rate. He has only received a total of 1200ml of IV solution. They report the patient's LOC continued to drop so they intubated him. His vital signs indicate a significant tachycardia, BP's are in the 100/50 range, he is being ventilated adequately. A foley catheter is in place but the primary care providers did not track output volumes and it appears the patient may have stopped making urine all together. You realize they have incorrectly managed the fluids. You need to reset the IV maintenance to what rate to correct per the Parkland Burn Formula?

a. 2000ml/hr
b. 3000ml/hr
c. 5000ml/hr
d. 6000ml/hr

54. Minimum ideal urinary output for our adult burn patient would be;
    a. 30-50ml/hr (~0.5ml/kg/hr)
    b. 1ml/kg/hr
    c. 2ml/kg/hr
    d. 3ml/kg/hr

55. Cyanide toxicity should be considered high on the list of possible complications with any confined space fire involving;
    a. natural fibers (cotton, wood, etc.)
    b. electrical equipment (wiring, conduit, etc.)
    c. petroleum products (gasoline, kerosene, Jet A, etc.)
    d. synthetic fibers (carpet, clothing, plastics, etc.)

56. The 'consensus formula' calculates hourly fluid resuscitation rates at;
    a. 4ml/kg/%BSA
    b. 2ml/kg/%BSA
    c. 1-3ml/kg/%BSA
    d. 2-4ml/kg/%BSA

57. Alkali metal burns;
    a. should be irrigated with copious amounts of water
    b. cause saponification
    c. cause coagulative necrosis
    d. should be neutralized with calcium gluconate

58. You have arrived at the scene of a hazardous materials spill and you realize this is a "fast-break scenario". As such you intercept a contaminated victim extricating himself from the hot zone. You should;
    a. direct them to return back to the spill and wait
    b. advise them to stop there, retreat and enlarge the hot zone
    c. perform a 2-stage decontamination
    d. run away, far far away

59. Electrical injuries will vary dependent upon all but which of the following?
    a. type of current (AC vs. DC)
    b. voltage
    c. amperage
    d. patient age

60. Myoglobinuria is commonly associated with all but which of the following?
    a. pneumothorax
    b. crush injury
    c. electrical injury
    d. burn injury

61. Which of the following is not a key component of myoglobinuria treatment?
    a. osmotic diuretics
    b. aggressive fluid resuscitation
    c. $NaHCO_3$ administration
    d. vasopressin/DDAVP administration

Scenario for triage questions:

Patient No. 1: Adult male presents limping towards you holding his open radial fracture. He's crying and yells, "Please help my daughter!" He has various abrasions and is covered in dirt. It appears he was ejected from the vehicle and has walked back to the vehicle's resting point.

Patient No. 2: Presents laying inside the upside down vehicle. Age 1 to 2 years. She is unresponsive and apneic even after manual airway maneuvers. Cap refill is delayed.

Patient No. 3: Presents on the ground next to the vehicle, it appears someone may have pulled him from the wreckage. He appears to be a 4 to 5 year old male. He is unresponsive, pupils are at ~5mm and unresponsive. He is breathing irregularly at 20-30 bpm, his cap refill is less than 2 seconds.

Patient No. 4 Presents face down with the lower body pinned under the vehicle. She is breathing spontaneously but agonally. She is unresponsive. Cap refill is more than two seconds. It appears that brain matter may be present at the right ear canal and nose, the frontal region of the skull appears grossly depressed.

Patient No. 5 Presents hanging from a seatbelt crying. He appears to be a 10-12 year old male who is obviously very upset and can't extricate himself from his seatbelt but is fighting to do so regardless of your efforts to calm him. He won't let you take vitals but he continues to cry, scream and kick. Skin color is flushed (he's upside down). There are no obvious injuries.

Patient No. 6 Presents in the back of the vehicle tangled in bedding and luggage. He appears to be the twin brother of patient No. 5. He is crying and guarding his right leg. The right leg demonstrates obvious angulation at mid-femur. He answers questions appropriately and repeatedly asks, "where's my mommy?" His cap refill is less than 2 seconds.

62. Per the START triage system, patient No. 3 would be;
    a. Priority I
    b. Delayed
    c. Green tagged
    d. Level II

63. Per the START triage system, you would triage the patients for treatment as;
    a. 2, 4, 3, 6, 1, 5
    b. 3, 6, 1, 5, 2, 4
    c. 4, 2, 3, 6, 1, 5
    d. 3, 6, 5, 1, 2, 4

64. The core components of START triage assessment are;
    a. respirations, perfusion and mentation
    b. airway, respirations and perfusion
    c. respirations, mentation and injuries compatible with life
    d. mentation, respirations and extrication time

65. Which of the following would be an abnormal finding on a complete blood count (CBC)?
    a. RBC 2.5
    b. Hgb 14.8
    c. Hct 44.1
    d. WBC 7.8

66. Which of the following would be an abnormal electrolyte level?
    a. Na 138
    b. K 5.8
    c. Cl 101
    d. TCO_2 23

67. The lab value, which identifies which portion of the blood is not plasma is the;
    a. hemoglobin (Hgb)
    b. hematocrit (Hct)
    c. differential "diff"
    d. red blood cell count (RBC)

68. A normal platelet count would be;
    a. 100-200K
    b. 150-250K
    c. 150-400K
    d. 200-600K

69. The predominant serum protein would be;
    a. globulin
    b. platelets
    c. dextrose
    d. albumin

70. Which of the following is an example of a isotonic solution?
    a. albumin
    b. D5 1/2 NS
    c. LR
    d. Hespan®

71. Blood loss should be replaced with crystalloid solutions in a ratio of;
    a. 1:1
    b. 2:1
    c. 3:1
    d. 4:1

72. Your patient has become hypovolemic and demonstrates signs of shock. Included with this you would anticipate;
    a. pre-renal failure
    b. renal failure
    c. post-renal failure
    d. none of the above

73. The clotting cascade can be triggered by the intrinsic or extrinsic pathways. The triggering mechanism, which initiates the extrinsic pathway, is;
    a. collagen exposure in the endothelium
    b. tissue thromboplastin release
    c. platelet damage
    d. thrombin release

74. Disseminated intravascular coagulopathy (DIC) is primarily a problem with;
    a. bleeding
    b. clotting
    c. platelet failure to function
    d. thrombin deactivation/desensitization

75. Your DIC patient would most likely demonstrate;
    a. low D-dimer
    b. high fibrinogen levels
    c. low PT and aPTT
    d. high fibrin split product (FSP) levels

76. Your focus when treating the DIC patient should be;
    a. replace clotting factors
    b. heparin administration
    c. ATIII replacement
    d. correcting underlying pathology

77. Your patient's H&H was 7 & 20. You have administered two units of packed red blood cells (PRBC's). You can reasonably assume his H&H should increase to;
    a. 10 & 23
    b. 8 & 21
    c. 9 & 22
    d. 9 & 26

78. The ABO antigens, which dictate blood type, are found;
    a. in the plasma
    b. on white blood cells (WBC's)
    c. on hemoglobin
    d. on red blood cells (RBC's)

79. You have administered 5 units of PRBC's in rapid succession during a trauma resuscitation. You should be considering;
    a. citrate toxicity
    b. anaphylactoid reaction
    c. methemoglobinemia
    d. hypervolemia

80. The corrective action for the problem suspected above would be;
    a. diphenhydramine and steroid administration
    b. diuresis and foley placement
    c. methylene blue administration
    d. calcium administration

81. A problem associated with rapid administration of PRBC's would be;
    a. hypernatremia
    b. hyperkalemia
    c. hyperbilirubinemia
    d. hypomagnesemia

82. Administration of large quantities of PRBC's may cause;
    a. Oxyhemoglobin dissociation curve shift to the right
    b. Oxyhemoglobin dissociation curve shift to the left
    c. Oxyhemoglobin dissociation curve shift upward
    d. Oxyhemoglobin dissociation curve shift downward

83. You have just begun administering a unit of PRBC's to your trauma patient. Approximately 3 minutes after initiating the blood you note a degree increase in temperature and the patient says their back hurts (new complaint). You should suspect what type of blood reaction?
    a. anaphylactic
    b. febrile
    c. hemolytic
    d. pyrogenic

84. You have been called to transport a patient who was given TPA for an AMI and has subsequently developed extensive bleeding. The best blood component needed to stop this reaction would be;
    a. PRBC's
    b. FFP
    c. Platelets
    d. Cryoprecipitate

85. The only solution which contains red blood cells and clotting factors would be;
    a. PRBC's
    b. whole blood
    c. FFP
    d. factor VIII concentrate

KEY & RATIONALE

TRAUMATIC INJURY & HEMATOLOGY

1. D Tracheal shift on external exam is commonly a late sign/pre-arrest sign of tension pneumothorax. While commonly mentioned in the list of signs for this pathology, it is seldom seen on external exam of the live patient.

2. C Kinking of the vena cava with acute preload decrease is the primary mechanism, which precipitates cardiovascular collapse. While collapse of the unaffected lung, hypoxia and myocardial compression does contribute it is the major venous compression that is primary.

3. B Tracheal shift may be seen in tension pneumothorax but seldom if ever with hemothorax. For the patient to acquire enough blood in the pleural space to displace the mediastinum they would need to hemorrhage more than their own blood volume, hence hypovolemic shock and arrest would precede any tracheal deviation.

4. D If the open chest defect is more than ⅓rd the size of the glottic opening, specifically between the vocal cords, air will preferentially entrain via the defect vs. the trachea.

5. C Immediate occlusion of the defect is indicated. To minimize trapped air in the pleural space at the time of occlusion, the dressing should be placed at end-exhalation, preferably a forced end-exhalation to allow venting of as much air as possible from the pleural space. If the patient were mechanically ventilated, optimal occlusion would occur at end-inhalation when pleural pressures would then be at their highest.

6. D Needle decompression may be safely performed at the following locations;

2nd intercostal, mid-clavicular
4th intercostal, anterior to mid-axillary
5th intercostal, anterior to mid-axillary and mid-scapular

Risks of improper site location include penetration of the subclavian artery/vein; liver, gallbladder, spleen, pancreatic injury and nerve paresthesias

7. D One should never wait for radiographic confirmation of a tension pneumothorax to treat. Identification is based on clinical findings alone. Absent breath sounds on one side are justification for decompression, especially if air transport is pending. Difficulty breathing alone can be from any number of causes to include simply splinting from pain at the chest injury site itself. Tracheal shift is much too late a sign to wait for prior to decompression.

8. B This is somewhat a "trick question". All answers are correct according to the American College of Surgeons ATLS curriculum and commonly practiced standards of care. However the text referred to for the exams this book addresses only cite option B as an appropriate site for placement.

9. D Positive pressure ventilation can and will typically exacerbate a pneumothorax very quickly, in as little as two breaths in some instances and risk of air embolism is drastically increased as well. While intubation typically includes positive pressure ventilation, the answer option did not. Tube thoracostomy is the definitive treatment for tension pneumothorax with needle decompression being the next best therapy (while admittedly a far 2nd). When managing the pneumothorax victim, avoid positive pressure ventilation as long as possible or until a functioning tube thoracostomy is placed.

10. D Cyanosis is commonly a late sign regardless the mechanism precipitating the hypoxia. It should also be remembered that the patient's hemoglobin typically needs to be greater than 5g/dl for cyanosis to be detectable so the pneumothorax victim with concomitant hypovolemic shock may never demonstrate cyanosis. Recall that any penetrating trauma between the angle of the jaw and the inguinal ligaments is considered thoracic until proven otherwise as angle of penetration/trajectory is indeterminable. Option B may seem attractive and would be true for a *pneumo*thorax, not hemothorax. Option C should be identified as a poor choice by it's use of the word "*all*" in the answer. Generally speaking, answers which include words like, "*always, all, completely, never*" or other 'absolute' terms are poor choices. While we try to identify rules to follow in medicine for ease of practice, it is the very dynamic nature of medicine that makes it so challenging a practice, thus "absolute" conditions seldom present themselves.

11. D Remember that along the inferior margin of each costal you will find a costal artery, costal vein and nerve bundle. Avoidance of the inferior costal margin is the very reason we are taught to 'walk' a needle over the top of the rib when performing a needle decompression. This same principle applies when placing a tube thoracostomy.

12. D Refer to the photo below for clarification. All chest tubes currently available in the United States have a radiopaque indicator line along their long axis. There is a break in this line at the most proximal port on the insertion end (seen mid-field of the right lung in the film below). This break should be visualized inside the pleural space (see below) to verify adequate depth of insertion. The scenario describes this break or the proximal port being at the entry point to the pleural space. The subcutaneous air described at the insertion point should also lead you to suspect that air is being entrained via the wound, into the tissue as it is sucked into the proximal port on the improperly placed chest tube itself.

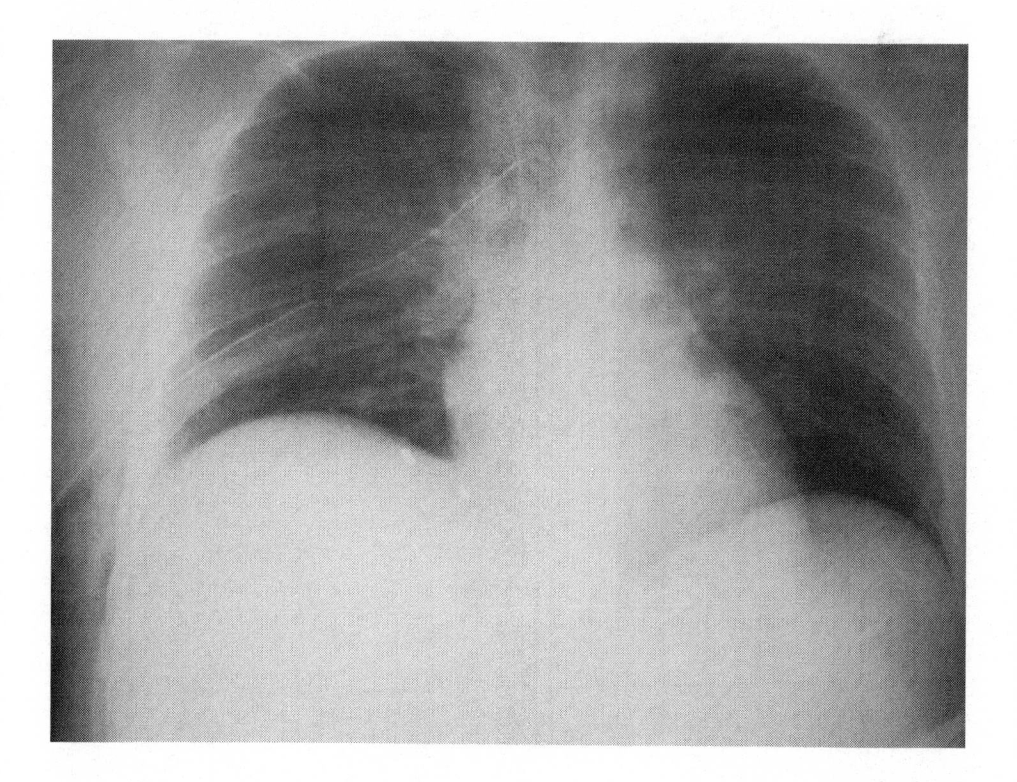

13. D The occlusive dressing should be applied at the point of highest intrapleural pressure and because this patient has already been intubated and positive pressure ventilated, that would be at end-inhalation.

14. B The purpose of taping the occlusive dressing on three sides is to permit a 'flapper-valve' mechanism to occur. Once the chest tube is in place you would like all air management from the pleural space to be managed via the chest tube, not the flapper valve. Thus, once the tube is placed, finish taping the fourth side of the occlusive dressing. Answer A is a neat little distracter and sounds very technical or official. If you've never heard of it, it *probably* is not a legitimate answer. Remember if you're an item writer you're task is to come up with incorrect answers that are plausible to the examinee not familiar with the material. If you are somewhat familiar with the material, you can typically identify the fabricated procedures, diseases and names pretty easily. If you were to turn this patient, you would ideally turn a thoracic injury patient injury side down to allow the uninjured lung to remain independent and also provide some tamponade of bleeding from the injured lung in the dependent position.

15. C While sandbag management of these injuries was once standard of care, it is no longer. Management of the flail would include the other three options and analgesia. Most important is likely the intubation with PEEP.

16. B Pulsus paradoxus or an exaggeration of the changes in pulse quality as the patient breaths is typical of early tamponade. Trousseau's sign is indicative of hypocalcemia/hyperphosphatemia. Jerisch-Bergen reflex involves bradyarrhythmias and ventricular arrhythmias associated with acute preload reduction and AV nodal hypoxia. Narrowing pulse pressures occur with tamponade, widening pulse pressures are typically associated with neurological insults and Cushing's Syndrome

TEST TIP
"pulsus paradoxus" ⇒ pericardial tamponade

17. A While counter-intuitive, the early management of cardiac tamponade is to 'push fluids' or augment preload. The initial pathology causes a diastolic failure or failure to fill the ventricles. While this management may commonly progress to exacerbating the tamponade there are no better options in the initial phases of therapy. Vasodilators and inotropes would not help and may likely harm the tamponade victim. Pericardiocentesis is the definitive therapy for acute tamponade but typically is not performed until the extreme/later stages.

TEST TIP
early pericardial tamponade ⇒ push fluids
late pericardial tamponade ⇒ pericardiocentesis

18. B While hypotension will likely be present during a tamponade, Beck's triad, the classic group of symptoms associated with tamponade, does not include it.

19. D Aortic rupture will result in harsh systolic murmur at all commonly auscultated cardiac sound locations. The harsher the murmur, typically the smaller the rupture. This general principle applies to murmurs associated with VSD's and ASD's in neonates/infants as well.

20. B A scaphoid abdomen will present if a significant portion of the bowel has migrated through the diaphragm into the thoracic cavity.

TEST TIP
"scaphoid abdomen" ⇒ diaphragmatic rupture/hernia

21. B The 'caval laceration' and suspected liver laceration will require emergent surgical intervention. During that intervention the vena cava will likely require 'cross-clamping'. During that period of time IV access both above and below the cross-clamp is necessary to provide fluid resuscitation in the event of a significant hemorrhage. Maintaining an adequate MAP is typically more important than the systolic BP, be cautious about assigning 'hard numbers' to systolic and diastolic pressures. While attaching a blood set and pressure transfuser is smart, attaching it to the central line is not as smart. You would be able to flow much more fluid via the 16ga in the AC versus the CVP line. Intubating may prove necessary at some point but it is not indicated based solely on the information provided. Intubating a fragile major vessel defect can also prove quite harmful if pressures are not monitored and managed very closely during the induction and intubation.

22. D All of the terms offered here are 'hints' for specific pathologies. These are all worth knowing for the exams as well. "Patchy infiltrates" is a common description of the interstitial fluid noted on many ARDS patient films. A "flattened diaphragm" is typical of any disease where air trapping is occurring, i.e- asthma and COPD. "Kerley's B lines" refer to the engorged vasculature that is seen in the pulmonary fields of a patient with congestive heart failure and/or pulmonary hypertension.

TEST TIP
"patchy infiltrates" ⇒ ARDS
"flattened diaphragm" ⇒ hyperinflation (asthma, COPD)
"Kerley's B lines" ⇒ CHF

23. B The 'classic' presentation of an aortic aneurysm is pain that is severe, across the back, constant, ripping or tearing in nature and without relieving factors. Option A is typical of an AMI. Option C is typical of pericarditis. Option D is just dumb.

24. A While the NG/OG tube may seem somewhat argumentative you must consider that this patient's bowel is in his chest where his lungs are. That means they're taking up valuable space that the lungs need for proper inflation and VQ matching. Removing any and all air from that bowel is essential to maximizing oxygenation and ventilation.

25. D This type of injury will subsequently precipitate injury to the bowel itself if not corrected in a timely manner. While the bowel is encarcerated it will likely become hypoactive and risk obstruction, diarrhea is unlikely.

26. D A constant irresolvable air leak with a good chest tube should lead you to suspect tracheobronchial disruption. Even if the mechanism of injury does not support that diagnosis, a traumatic intubation performed prior to your arrival may be all that is needed. While an esophageal perforation can contribute to a hemo/pneumothorax, air passage would be completely passive and very limited, something a chest tube could easily clear and stay ahead of (assuming of course you didn't 'tube the goose' and pressure ventilate their esophagus).

TEST TIP
"continuous air leak" via a chest tube ⇒ tracheobronchial disruption

27. D Placing an endotracheal tube with a cuff beyond the disruption is the definitive therapy for a tracheobronchial disruption. In the surgical suite a dual lumen tube or a Bronchial Blocker is placed and fiber optics used to verify placement of the cuff beyond the injury. In the field one may attempt to gently place the ETT in a deep position, re-inflate and verify ventilation without the air leak. Obviously adjustment for tidal volume may need to be made to prevent injury while performing one lung ventilation (OLV). If attempting to perform OLV, ideally you would like to place the patient into a lateral recumbent position with the ventilated lung in the dependent position. This ensures optimal VQ matching in a suboptimal situation (consider West zones). Also consider some patients will not be able to tolerate OLV very well. A prehospital strategy one might employ would be insertion of a french ET suction catheter down the trachea next to the ETT. Supplying O_2 via this suction catheter, (it's tip proximal to the ETT cuff) may allow enough passive O_2 supply to the 'downed lung' to keep saturations up and prevent total de-recruitment.

28. D It may surprise you that an OG/NG tube should be placed here. You must remember that the patient will be bleeding into the esophagus and subsequently collecting in the stomach. Blood is very irritating to the stomach and will surely prompt vomiting. Vomiting is actually the most common cause of esophageal perforations due to its corrosive nature and high pressure. A patient vomiting into their chest cavity is not good, hence you should carefully place a gastric tube and keep the stomach evacuated. Answers B and C hopefully are intuitive. Answer D may sound fine, keep in mind that narcotics again promote vomiting, especially morphine. Morphine also is a fairly potent vasodilator due to it's histamine release characteristics and this may contribute to shock as well as continued hemorrhage, not to mention potential respiratory depression.

29. C Sir Isaac Newton's first law of motion states that an object in motion will remain in motion and an object at rest will remain at rest unless acted upon by an outside force. Newton's second law is simply Force (F) = mass (m) X acceleration (a) [F=ma]. Newton's third law is stated in the question. Einstein's theory of relativity has not been identified as relevant to EMS as of yet. Yes, you need to know Newton's three laws for the exams.

TEST TIP

Newton's 1st Law: "An object in motion will remain in motion, an object at rest...."

Newton's 2nd Law: "F = ma"

Newton's 3rd Law: "For every action there is an equal and opposite reaction"

30. B Remember that in the 'up and over' injury pattern the patient's body has attempted to go 'up and over' the steering wheel. As such you should anticipate head injury from impact with the windshield, c-spine injury from the same, chest injuries from impact with the steering wheel and in the event a lap belt was worn, pelvic injuries may need to be ruled out. The second common injury pattern is the 'down and under'. With this pattern you should anticipate the patient slid down and under the steering wheel, commonly striking the knees against the dash of the car. This precipitates ankle fractures, tib-fib fractures, femoral fractures, acetabular fractures with hip displacement, lower back injury and if a lap belt was worn, consider liver, spleen and pancreatic injuries as well.

31. A All of the signs listed are commonly associated with LeFort fractures with periorbital ecchymosis or "raccoon's eyes" more commonly associated with a basilar skull fracture. Epistaxis presents almost without exception while the other effects may not be as common. Interestingly enough, LeFort fractures all cross various branches of the 5th cranial nerve, the trigeminal nerve, and as such trigger the trismus so commonly seen with these maxillofacial injuries.

32. B Generally speaking, anyone with a LeFort facial fracture has a very high chance of concomitant basilar skull fracture. NG tube penetration into the cranial vault is well documented and completely unpreventable short of simply not placing the tube initially. A and C are likely very intuitive. D should be placed to monitor fluid status as this patient may likely have a head injury requiring further diuretic management, surgical management and will almost undoubtedly spend substantial time in an ICU setting.

TEST TIP

Basilar skull fracture/maxillo facial fractures ⇒ avoid placing anything in the nose

33. C Only a paralytic will stop the trismus and permit orotracheal intubation. Benzodiazepines and opioids may make the patient more comfortable, less anxious and more forgetful about their situation but they will not break the muscle tension in the jaw. This scenario will require either paralytic assisted intubation or a surgical airway.

34. D The two spinal injury sites offered are very common in rear-end collisions. The C2 fracture is typical in vehicles with the headrest set too low/removed and is commonly referred to as a 'hangman's fracture'. Ankle fractures result from the foot slipping from the brake pedal during the collision with enough force to twist and fracture the articulation at the ankle.

35. B Up and over refers to the rider striking an immovable object flying up and over the handlebars, typically injuring the head, neck and chest, commonly arm and shoulder injuries are also found as they attempt to break the fall. The 'lay it down' pattern refers to the rider attempting to stop the bike goes into a sideways slide and effectively does just that; he 'lays the bike down'. Injuries commonly include trapped leg fractures, pelvic fractures and road rash. Some texts will also cite rib fractures resulting from the 'trapped arm' on the dependent side. If you happen to be at a motocross/super-cross race you'll likely see option C more commonly, they're sure more fun to watch, but not used on the exams.

36. C Rollovers kill more people each year than any other type of motor vehicle crash. The advent of passive restraint systems have significant reduced trauma from head on and side impact collisions.

37. A C-spine must be managed simultaneously with the earliest stages of airway management. Everyone reading this with any significant trauma background will find this a 'no brainer' but our colleagues coming from a predominantly medical background may find this new territory.

38. C We must lose a third or more of our blood volume before a drop in blood pressure is typically detected. This is why waiting for this blood pressure drop places you well behind the curve in treating the trauma victim.

39. B For rapid infusion you want the catheter with the biggest bore and shortest length. The central line may seem enticing but it's length makes it much slower. Be sure you are not confusing the catheter described with a "cordis" or introducer cath sometimes placed with the central line laced through it. An introducer catheter is ideal for fluid resuscitation as it is typically an 8fr or 9fr which translates to approximately a 2.5-3ga catheter. Remember we never infuse anything via an arterial catheter, they are only for monitoring and blood draws.

40. C In the acute phase we must use a crystalloid that is ideally isotonic. The only such fluid listed is LR. All dextrose containing solutions will be hypotonic and rapidly third space after a single circulation time. Albumin, a colloid, will remain in the vascular space somewhat longer and may pull fluid from the interstitial space. While this may sound good in theory it is a poor choice in the earliest stages of resuscitation and research has questioned it's appropriateness even in late stage resuscitations.

41. A The increase in heart rate is compensatory. This means he/she needs help. Altered level of consciousness may be a nice confirmation of shock but it can also simply be a sign of closed head injury, toxins, drug ingestion, etc. Decreases in oxygen saturation are almost uniformly a late sign of their etiology, whatever it may be.

42. B Reassess, reassess, reassess. Vital signs in and of themselves are not very useful. Trends of vitals are very useful, hence reassessment is critical. Obviously constant assessment of hemostasis, hemodynamics, oxygenation and ventilation are equally critical.

43. B Farming equipment is heavier but typically has a *higher* center of gravity thus making them prone to rollover type accidents. Fertilizer and fuel spills are common as well as the operator frequently being pinned by the heavy equipment due to lack of protective cage or roll-bar. Medical care is commonly delayed, as the patient is not missed until he doesn't "come in for dinner", meanwhile he's been laying under a tractor since 6:00am when he started his daily work routine.

44. A Typically cervical and lumbar axial loading injuries are seen. The basic premise of the syndrome is that the adult leaps from a substantial height attempting to land on their feet. They fracture bilateral calcaneus (heels), tib-fib's, femurs, pelvis; axial load their lumbar and cervical spines (points where the spine is least supported then fall back with outstretched arms and receive Colles' fractures to wrists.

45. C The 'textbook' answer is that females typically hold the knife with the blade extending from the heel of the clenched fist and stab downward with a "Psycho killer" type motion. Men on the other hand stab upward, holding the knife with the blade extending from the thumb side of the clenched fist just like every "bad guy knife-fighter" we watch on TV. Yeah....right.

46. B Per the exam references, "high velocity" is any weapon firing a projectile faster than 2000 fps which would limit them to mainly long guns (aka- rifles.)

TEST TIP
"High velocity weapon" ⇒ >2000fps

47. D Type, hand gun (pistol) vs. long gun (rifle). Bullet type; slug vs. hollow point. Distance; amount of energy still present upon impact, are all important. Manufacturer is irrelevant. Smith & Wesson's bullets are no more or less lethal than Federal's or Winchester's or Remington's....

48. C There are three injuries involved with blasts. The primary injuries are those caused by the initial blast wave striking the body and compressing air filled spaces causing ruptures. The secondary injuries are caused by debris or shrapnel from the explosive device penetrating or striking the body. The tertiary injuries are those caused by the body being knocked to the ground from the blast itself with subsequent blunt trauma resulting.

TEST TIP
Blasts cause three injuries; primary, secondary and tertiary

49. A Recall your rule of nine's and also remember that you only count areas with 2nd and 3rd degree burns. Thus calculating 9% for the entire face and head, 18% for the upper chest and back, 9% for each complete arm and hand and you add them up to 45% total BSA.

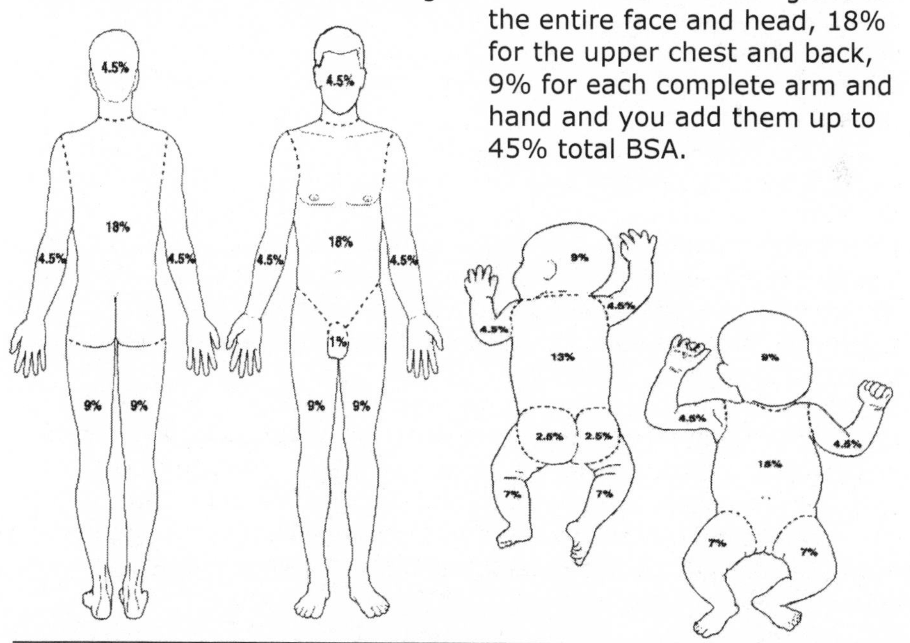

50. D The Parkland Burn formula is still referenced on these exams. Per that formula, fluid resuscitation is calculated at 4ml/kg/%BSA for total resuscitation volume. One half of that total volume should be given in the first eight hours after the injury, the remaining half over the subsequent 16 hours. It should be remembered that this formula provides a 'maintenance rate' only. If you arrive on scene and the patient presents in hypovolemic shock you should fluid resuscitate like any other shock patient and once appropriately resuscitated, then calculate where you are in the maintenance volume and readjust your maintenance rate from there.

TEST TIP
Parkland Burn Formula
Only calculate %BSA with 2nd and 3rd degree burns
4ml/kg/%BSA burned = total fluid resuscitation volume for the 1st 24 hours
½ total resuscitation volume is given in the 1st 8 hours post injury

51. C Cyanide and carbon monoxide poisonings should be high on your list of suspected complications with anyone involved in a confined space fire. The fact this patient was 'cooking Meth' tells you that he's likely under the influence of the drug and may be experiencing the associated delusions, hallucinations and/or paranoia. Extreme caution must be exercised with this subset of patients regarding personal safety.

52. D Your initial Parkland calculation should have revealed a total of 14.4 liters for the first 24 hours. One half of that should be administered in the first eight hours. Thus 7.2 liters over eight hours would require an initial IV maintenance rate of 900ml/hr. (7200ml/8hrs)

53. D You are at hour seven of the first eight hours post injury. Based on the initial calculated maintenance of 900ml/hr, the patient should receive a total of 7200ml of IV fluid from the time of injury until one hour from now. They have only received 1200ml's thus far. You need to catch up by infusing 6000ml's in the next hour. Obviously this could be problematic if there is concomitant disease such as CAD, CHF, etc. You would also need to get some "big lines" if you're going to pull this off. This isn't ideal but it is correct per the burn management sections of the texts. (Please evaluate carefully before infusing 6000ml of crystalloid into an actual patient in an hour, this *can* be quite hazardous.)

54. A An easy way to remember ideal urine output is;
0.5ml/kg/hr for adults
1.0ml/kg/hr for peds
2.0ml/kg/hr for neonates

TEST TIP
'Normal' Urine Output
0.5ml/kg/hr Adult
1ml/kg/hr Peds
2ml/kg/hr Neos

55. D Synthetics like nylon and plastic are wonderful cyanide generators as they are incompletely combusted. Look for cyanide poisoning with house fires, especially when furniture and *carpet* are involved.

TEST TIP
Burning carpet ⇒ cyanide poisoning

56. D The consensus formula was derived from an attempt to find a mutual ground between the Brooke Burn Formula (2ml/kg/%BSA, derived from Brooke Army Medical Center, the Army's burn center) and the Parkland Burn Formula (4ml/kg/%BSA, derived from Parkland Burn Center). Yes, it's on the test.

TEST TIP
Consensus Burn Formula 2-4ml/kg/%BSA burned

57. B Alkali metal burns are commonly considered the worst because they effectively denature lipids into soap (*saponification*). Alkali's are very reactive with water thus irrigating with copious amounts will only antagonize the reaction and potentially cause ignition and fire. (This makes it harder to treat the patient because they keep rolling around in the dirt screaming, "Fire! Fire!" This will also scare the bajeezus out of most pilots if it occurs in the aircraft.) Most acid burns cause *coagulative necrosis*, a point worth noting for the exams. Calcium gluconate may be used to neutralize only hydrofluoric acid burns, all other acids are simply irrigated with water (dilution, dilution, dilution).

TEST TIP
Alkali burn ⇒ saponification
Acid burn ⇒ coagulative necrosis
"The solution to pollution is dilution"

58. C The long-time traditional management of a hazardous material spill involved a long staging and isolation process followed by an equally involved decontamination procedure. Time has shown that this simply doesn't occur initially. Initially we see a "fast-break scenario" where contaminated victims are attempting to leave the hot-zone in an effort to save themselves. EMS providers may encounter these victims prior to Haz-Mat Team arrival and as such should perform a *two-step decontamination*. This procedure involves directing the victim to stop, undress and be washed with clean water and soap if available. They should then step out of the run-off and the wash be repeated a second time.

59. D Patient age in and of itself is irrelevant. Assuming concomitant disease associated with age will complicate the injury is very intuitive but you are reading into the question.

60. A Myoglobinuria or myoglobin in the urine is frequently associated with any pathology that results in muscle breakdown (rhabdomyolysis). Recall that myoglobin is an iron containing pigment found in "slow-twitch" or red muscle and acts as an oxygen reservoir in said tissue. Myoglobinuria must be anticipated and treated aggressively with electrical burns, thermal burns and any form of massive trauma or crush injury.

61. D Treatment of myoglobinuria should focus on driving the kidneys to produce urine. This commonly involves aggressive fluid administration and osmotic diuresis. Administration of the $NaHCO_3$ creates an alkaline environment, which inhibits the myoglobinuria from binding to other proteins making it easier to pass through the glomerulus of the kidney. Vasopressin or DDAVP would cause fluid retention, which is not desirable here.

TEST TIP

Myoglobinuria ⇒ push fluids, $NaHCO_3$ and mannitol

62. A START triage would categorize him as a "Red tag" also termed "Priority I" or "Level I". The exam may use any or all of these terms so you need to know the following;

Black tag	= Priority 0	= Deceased
Red tag	= Priority I	= Level I
Yellow tag	= Priority II	= Level II
Green tag	= Priority III	= Level III

63. B It is always difficult to triage small children because of our natural instinct to protect and help the young and weak. Patient No. 1 is a "walking wounded" and is assigned a "Green tag". Patient No. 2 is not breathing even with manual airway maneuvers. This fact alone gets her "black tagged" on initial triage. Patient No. 3 has an airway, spontaneous but altered respirations and normal cap refill. The altered respirations and mentation get him a "Red tag" priority. Patient No. 4 show's brain matter exposed, which considered along with her other presenting factors also assigns her a "black tag". While being pinned isn't a factor in the START system you would have to consider that fact when faced with six potential patients and only you and your partner to render aid. Patient No. 5 is one upset little boy but he's basically uninjured on the initial assessment and thus "walking wounded" or "Green tagged". Patient No. 6 has a good airway, normal breathing and cap refill. This would get him "Green tagged" but his obvious femur fracture prevents him from being a "walking wounded" and thus escalates him to a "Yellow tag" assignment. To summarize;

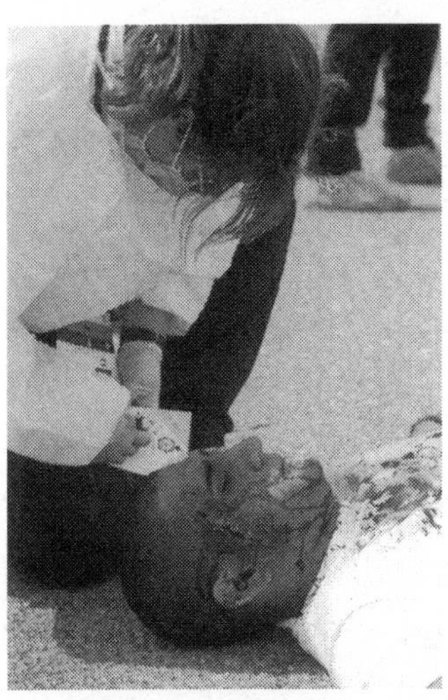

Patient No. 1 - Green tagged - Priority III - Walking wounded
Patient No. 2 - Black tagged - Priority 0 - Dead on Scene
Patient No. 3 - Red tagged - Priority I - Immediate
Patient No. 4 - Black tagged - Priority 0 - Dead on Scene
Patient No. 5 - Green tagged - Priority III - Walking wounded
Patient No. 6 - Yellow tagged - Priority II - Delayed

64. A When performing START triage remember the mnemonic, "When your mind starts racing, remember your RPM's" Respirations, Perfusion and Mentation. Assess respirations as absent, normal or abnormal. Perfusion is assessed as central core cap refill being normal, delayed or absent. Mentation is assessed as normal, abnormal or absent (injuries incompatible with life). So the basic routine on a mass casualty incident is upon arrival you say, "Everyone who can walk, go to location X for evaluation." All of those people are typically given Green tags. The remainder of patients laying around then get a 15 second assessment of respirations, perfusion and mentation being given a Red Tag, Yellow Tag or Black Tag priority. You will be required to triage a group of patients in a manner similar to this on your exam.

65. A You will need to know some basic ranges of lab values for these exams. Basic lab values/ranges you should know include the following;

TEST TIP
Know the labs below

CBC

RBC	5	*These are easy to remember if you simply memorize the RBC value and multiple times 3 for the Hgb and again for Hct.*
Hgb	15	
Hct	45	
WBC	5-10K	
Plt	150-400K	

CHEM 7

Na^+	135-145	
K^+	3.5-5	
Cl^-	95-107	
TCO_2	22-26	(Basically a serum representation of the HCO_3 you get on an ABG)
BUN	10	
Cr	1	
Gluc	70-110	

66. B Normal potassium is 3.5-5.

67. B The hematocrit (Hct) identifies how much of the blood is particulate or "stuff" like RBC's, WBC's, platelets, etc... Normally about 45% of your blood is "stuff" and not plasma. In case you haven't already figured this out, the "industry standard *lingo*" of "H&H" always refers to Hgb & Hct, in that order.

68. C Refer to rationale for question 65.

69. D The primary serum protein is albumin with globulin a distant second. RBC's are made of protein and thus participate in binding medications and acids as well.

70. C Isotonic solutions are solutions which have the same tonicity as blood, crystalloids such as normal saline (NS) and lactated ringers (LR) (aka-Ringer's Lactate). Any dextrose solution will be hypotonic after a single pass in the circulatory system unless it's a hypertonic solution such as D50W, which will also become hypotonic after multiple passes. Albumin and *Hespan*® are both hypertonic colloidal solutions as they contain protein.

71. C This replacement ratio is required as ⅔**rds** of crystalloid will 'third-space' to the interstitium rapidly.

72. A The term "pre-renal" failure suggests that the renal system is not producing urine secondary to a failure "before" the kidneys. Thus, failure to perfuse the kidneys with adequate pressure, like in shock, precipitates "pre-renal" failure. Some have said, "pre-renal failure is renal success" emphasizing that the kidney's failure to diurese fluid during a shock state is exactly what they *should* do for the patient to survive ultimately. Renal failure suggests that the failure to perform diuresis is due to pathology in the kidney itself or the nephrons. Post-renal failure is identified as a failure to eliminate urine due to pathology "after the kidneys", for example, a urethral blockage.

73. B The extrinsic pathway is triggered by tissue thromboplastin, which is released any time tissue is damaged. Tissue thromboplastin is found at highest concentration in the brain and the placenta. This is part of the reason we see such high incidence of DIC with head injuries and placental injuries (previa and abruptio) Collagen exposure in the endothelium is the primary trigger for the intrinsic side of the cascade.

74. B Refer to the name, "coagulopathy". It is very intuitive to think of DIC as a patient bleeding to death and indeed, bleeding becomes problematic. It is however the process of clotting and incomplete clot breakdown that actually causes the systemic death as small clot fragments collect in the microvasculature of the brain, liver, spleen, kidneys and lungs limiting blood flow with subsequent organ failure

75. D The typical DIC victim will demonstrate a high D-dimer, low fibrinogen levels (as the clotting factors are used up), a high PT and aPTT (they can't stop bleeding as clotting factors are used up) and high FSP's (fibrin split products, aka- FDP's, fibrin degradation products), which are the remnants of clots breaking up.

76. D The theories behind treating DIC are varied and none have proven better or worse than the others to date. Only one strategy is consistent amongst them all. Correct the underlying pathology. If the DIC is secondary to massive head trauma, we need to correct that ASAP for the DIC to correct, unfortunately the patient commonly won't survive the DIC long enough to receive definitive care for the underlying pathology.

77. D You can assume that the H&H will increase by a factor of 1 & 3 respectively with each unit of PRBC's administered.

TEST TIP
1 unit PRBC's raises H&H by 1&3 (respectively)

78. D The ABO antigens themselves are on the patient's RBC's. Remember your universal donor is O negative and your universal recipient is AB positive.

TEST TIP
Universal donor : O negative
Universal recipient: AB positive

79. A PRBC's are stabilized and preserved somewhat with citrate. Citrate however does chelate ionized calcium and degrades 2-3DPG. While the 2-3DPG is not recoverable, the calcium chelation can become problematic if the patient's liver function is abnormal or the administration rate overwhelms the liver's natural ability to manage citrate. If blood administration is exceeding 1ml/kg/min or approximately one unit every 5 minutes you should consider citrate toxicity probable and administer calcium. Remember that calcium is essential at multiple points in the clotting cascade for proper blood coagulation, not to mention all of the neuromuscular components that calcium affects. Technically, citrate toxicity is seen more with FFP administration than PRBC administration. Remember that administration of calcium is not benign, ideally serum ionized calcium levels are preferred for guiding therapy.

80. D See rationale for question 79.

81. B Potassium levels climb very high in banked PRBC's as some of the RBC's die and lyse. Potassium levels in banked blood can climb as high as 17-24mEq/L. Rapid administration of blood should cue you to monitor the EKG for signs of hyperkalemia. In the event this is suspected you must stop blood administration and administer CaCl, $NaHCO_3$, insulin and dextrose as needed.

82. B Recall that the Oxyhemoglobin dissociation curve only shifts left and right. Causes of **L**eft shift include; a**L**kalosis, **L**ow CO_2, **L**ow acid, **L**ow 2-3DPG, **L**ow temp and **L**ots of CO. Administering large volumes of banked blood may cause citrate toxicity and lowering of intrinsic 2-3DPG levels (**L**ow 2-3DPG.) As such you should anticipate the RBC's would want to ho**L**d the oxygen and not release it to the tissues as the curve shifts **L**eft.

83. C The classic hemolytic reaction begins within a few minutes of initiating a transfusion of blood causing hyperthermia and back pain. This needs to be addressed immediately by stopping the blood and treating any symptoms that present. 30-50% of hemolytic reactions proceed to DIC and **die**. Support of pressure with vasopressors and/or fluids is essential. Save the blood for further analysis by the blood bank. Febrile reactions are similar but typically require 30-90 minutes for onset of symptoms. Anaphylactic reactions are typical of any other anaphylactic reaction with typically rapid onset of pruritis, hives, respiratory distress, etc.

84. D PRBC's may prove necessary eventually but they will only be used to replace the lost red blood cells and will not stop the bleeding. FFP can help stop the bleeding as fresh frozen plasma is primarily clotting factors from donor blood. Platelets again may be helpful but will only help to replace that which has been lost by the bleeding. Granted without platelets you cannot stop the bleeding but simply overwhelming the patient with mass amounts of platelets will not stop bleeding in this scenario. Cryoprecipitate is a concentrated solution of factor VIII:C, XIII, vonWillebrand factor and most importantly, fibrinogen. Cryo is the most important agent with a TPA induced hemorrhage. In this scenario the patient would likely receive all of the components mentioned but Cryo has the best chance of stopping the bleeding.

85. B See rationale for question 84. Whole blood is blood that has been only 'cleansed' or washed and not separated into it's respective components. Whole blood is infrequently used in medicine today.

GENERAL MEDICAL

1. When receiving the patient with diabetic ketoacidosis (DKA) you would anticipate the potassium level typically to be;
   a. elevated
   b. low
   c. normal
   d. extremely elevated

2. Diabetic ketoacidosis is primarily a problem with;
   a. elevated blood sugar
   b. inadequate insulin receptors
   c. inadequate insulin or insulin resistance
   d. low blood sugar

3. You have just arrived to transport a 32yo, 5'6", 90kg female patient diagnosed with DKA. The patient has reportedly been breathing 40 breaths per minute for the last two hours in the Emergency Department. Upon your examination you determine that the patient is approaching exhaustion and ABG's indicate the following: pH 7.22, $PaCO_2$ 21, PaO_2 289, HCO_3 16 on NRB mask at 15lpm. You and your partner decide to intubate the patient for transport. Which of the four plans below would be the best to start with?
   a. continue sedation and paralysis of the patient after ETT placement, then set vent to; Volume Control, SIMV, rate 12, FiO_2 0.6, Vt 900ml, PEEP 5cm
   b. continue sedation and paralysis of the patient after ETT placement, then set vent to; Volume Control, AC, rate 12, FiO_2 0.8, Vt 600ml, PEEP 0
   c. continue sedation of the patient after ETT placement, then set vent to; Volume Control, AC, rate 20, FiO_2 0.6, Vt 900ml, PEEP 5cm
   d. continue sedation of the patient after ETT placement, then set vent to; Volume Control, AC, rate 20, FiO_2 0.6, Vt 500ml, PEEP 0

4. A key component used in the management of both DKA and hyperglycemic, hyperosmolar, non-ketotic coma (HHNC) is;
   a. aggressive fluid hydration
   b. rapid decrease of blood glucose level using insulin
   c. paralysis with normalized ventilatory control
   d. aggressive correction of acidosis using respiratory management and bicarbonate administration.

5. A common rationale cited for the reason HHNC patients have much higher blood glucose levels than DKA patients is;
   a. HHNC patients are elderly and thus LOC changes are not noted as soon
   b. HHNC patients are typically on TPN a major contributor to causing HHNC
   c. HHNC patients are typically known diabetics and receive some insulin in their daily care
   d. HHNC patients produce some insulin therefore their symptoms are not noticed until later in the disease

6. Which of the following choices would be best suited to treating the patient with diabetes insipidus?
   a. aggressive glucose control using insulin
   b. aggressive acidosis correction using respiratory control and bicarbonate administration
   c. aggressive diuresis using loop and osmotic diuretics
   d. aggressive fluid administration with DDAVP

7. Diabetes insipidus results secondary to;
   a. inadequate insulin production or insulin resistance
   b. inadequate ADH production
   c. over stimulation of aldosterone receptors
   d. acute dehydration

8. Which lab finding would be most associated with diabetes insipidus?
   a. elevated blood glucose
   b. relative hyperkalemia
   c. relative hypocalcemia
   d. urinary hypo-osmolality

9. The most common type of diabetes insipidus is;
   a. nephrogenic
   b. pathogenomic
   c. hereditary
   d. central neurogenic

10. Syndrome of inappropriate antidiuretic hormone (SIADH) is commonly reflected in lab studies by;
    a. hyperkalemia
    b. hypercalcemia (dilutional)
    c. hyponatremia (dilutional)
    d. hypoglycemia

11. You have been requested to transport a patient for respiratory distress. Upon arrival you are informed that the patient was diagnosed with small cell carcinoma of the lung three months ago and has progressively deteriorated. All of the following would be common complications of this diagnosis except;
    a. seizures
    b. pulmonary embolism
    c. pancreatitis
    d. SIADH

12. You are transporting a patient diagnosed with SIADH. Which of the following would be an appropriate therapy during transport?
    a. aggressive hydration or fluid resuscitation
    b. furosemide administration
    c. administration of vasopressin
    d. administration of aldosterone substitute

13. You are transporting a 57-year-old male with a history of chronic alcoholism. He presented to the emergency department unconscious with bloody vomitus. He is being transported with a diagnosis of acute GI hemorrhage. Appropriate therapies would include all but which of the following;
    a. OG/NG tube insertion
    b. administration of PRBC's
    c. administration of octreotide (*Sandostatin*®)
    d. all of the above are appropriate

14. Vasopressin may be administered to the acute GI bleed patient in an effort to vasoconstrict the upper GI system and limit hemorrhage. One should consider which of the following conditions a relative contraindication to it's use.
    a. pancreatitis
    b. cancer
    c. renal failure
    d. coronary artery disease

15. Your patient is a 17-year-old female who presents with anxiety and crying. She has a chief complaint of "being stressed" and palpitations.
    Vital signs are; HR 156 and irregular, RR 26, BP 118/78, SpO_2 99%, Temp 99.5°F. EKG demonstrates atrial fibrillation with uncontrolled ventricular response. Based on the information provided, the most likely diagnosis from the choices below is;
    a. psychogenic hyperventilation syndrome
    b. anxiety attack complicated by early pregnancy
    c. Cushing's syndrome
    d. Grave's disease

16. Your 56-year-old migrant female patient has taken 100mg hydrocortisone daily for the past two years per her doctor in Mexico for treatment of a disease, which you believe from the translated history, is lupus erythematosus. She was unable to purchase the medications recently and stopped taking them. Today she had two syncopal episodes followed by a seizure. Initial responders found her with a heart rate of 64, BP 72/32 and breathing at a rate of 10. Other tests indicated hypoglycemia which was treated appropriately. They initiated a dopamine drip after a two liter crystalloid resuscitation but nothing seems to bring her blood pressure up. You would anticipate which of the following disorders is most likely present?
    a. Grave's disease
    b. Cushing's syndrome
    c. Addison's disease
    d. Myxedema coma

17. Your 21-year-old female patient is reporting a recent history of weight loss, "hot flashes", anxiety and "heart racing". Today she complains of dyspnea and weakness. Vital signs indicate; HR 128, RR 24, BP 96/44, SpO_2 98%. Based on the information provided, the best diagnosis from the choices offered is;
    a. hyperthyroidism
    b. hypothyroidism
    c. hyperaldosteronism
    c. adrenal insufficiency

18. Your patient presents with a traumatic hand injury and will be transported to a tertiary care facility for surgery. The patient reports a two-year history of taking dexamethasone (*Decadron®*) for "allergies" which he obtains at a pharmacy in a Mexican border town nearby. He has notable bruising on his arms, a rounded face and obese habitus. You would anticipate all of the following findings except;
    a. abdominal striae
    b. buffalo hump
    c. weakness or fatigue
    d. hypoglycemia

19. Your patient presents with acute respiratory distress, extremely elevated white blood cell count, elevated bands, positive Cullen's sign and upper abdominal pain. His history includes alcoholism with frequent nausea and vomiting and poor medical care. Chest X-ray demonstrates diffuse patchy infiltrates primarily in the left lower fields with elevation of the left diaphragm. His abdominal pain is constant, without relieving factors and severe. Which of the following is the most likely primary diagnosis?
    a. acute appendicitis
    b. acute GI hemorrhage
    c. acute pancreatitis
    d. acute hepatitis

20. You have been called to a scene for a chief complaint of "man down". On arrival you find an apparently homeless male, 40 to 50 years of age with an obvious odor of alcohol about him and his clothing. He is somewhat rousable but you notice when he attempts to move his muscle movements are jerky and exaggerated with a notable coarse muscle flapping motion of his arms. This may be a good indicator of;
    a. alcohol poisoning
    b. methemoglobinemia
    c. antifreeze ingestion
    d. hepatic encephalopathy

21. Your patient has a complaint consistent with Kehr's sign. She reports she fell on the back frame of a lawn chair when she tripped striking her left lower rib region. Which organ is most likely injured?
    a. liver
    b. spleen
    c. gallbladder
    d. stomach

22. The patient is demonstrating a positive Murphy's sign. This would suggest a problem with his;
    a. liver
    b. stomach
    c. gallbladder
    d. spleen

23. The patient has been diagnosed with meningitis. You would likely find which of the following signs?
    a. Brudzinski's
    b. Levine's
    c. Grey-Turner's
    c. Cullen's

24. The patient is presenting with classic signs and symptoms of an acute myocardial infarction. This would include;
    a. Brudzinski's sign
    b. Levine's sign
    c. Grey-Turner's sign
    c. Cullen's sign

25. Your patient responds to extension of the hamstring with a desire to flex the head abducting the chin from the chest. This is;
    a. Brudzinski's sign
    b. Kernig's sign
    c. Chvostek's sign
    d. Grey-Turner's sign

KEY & RATIONALE

GENERAL MEDICAL

1. B Remember that as the DKA patient develops higher blood glucose, the increased serum osmolality will pull fluids from the interstitium to the intravascular space. The patient will in turn then diurese the excess fluid, taking potassium with it. As a general rule, a DKA patient is commonly potassium depleted. If their labs indicate a 'normal' potassium, you need to factor how much their acidosis has shifted their potassium from their intracellular to intravascular space. Recall that for every change in pH of 0.1 you will see a change in potassium of 0.6 the opposite direction. I.e.- your patient's pH decreases from a perfect 7.40 to 7.30, their perfect potassium of 4.0 will increase to 4.6. Thus as your DKA patient becomes more acidotic, more potassium is shifted from the cells to the bloodstream which will mask (as lab studies are derived from blood samples) that in reality, the entire body is becoming depleted of potassium as they continue to lose it with polyuria.

2. C DKA is a problem with lack of insulin or ineffective insulin. DKA patients typically do not manufacture insulin vs. HHNC patients make insulin, just not enough.

3. D Now I realize these settings are not ideal and you may not like any of the choices. The point here is to identify a couple key points.

First, paralysis. If you read this scenario, or possibly lived this scenario, many of you may have thought, "I need to paralyze this guy. I want him to breath when **I** say to breath and **only** when **I** say to breath. I must control everything." If your initial response came from deep down in your gut as a growling, "PARALYZE!" and it felt good to say it.... it's probably wrong. Very typical response considering the personality types involved here though. You must ask yourself, "Why is this patient breathing 40 breaths per minute?" The answer is to compensate for the acute metabolic acidosis they are experiencing. Without that respiratory alkalosis compensation their acidosis would become rampant. If you paralyze this patient you remove their ability to compensate completely. Not a good idea.

Second SIMV vs. AC. SIMV is typically considered a safer form of ventilation. Yes, in the patient that is not closely and competently monitored/managed this is especially true. We are trying to continue to provide the compensation they need along with removing the workload of actual breathing.

AC or assist control will provide the patient a full breath with every inspiratory effort with a minimum rate based on the frequency we have set. Thus setting an AC mode with an adequate rate allows the patient to get a full assisted breath each time they want to breath over and above what the vent has been set to provide (assuming you didn't paralyze them). So per the answer selection, this patient will receive 20 breaths regardless plus a full breath each time they take one over and above that. This is a nice technique to take workload off of a patient that is compensating such as in this scenario. The FiO_2 of 0.6 should be adequate. The current therapy is providing an FiO_2 of around 0.6 with a resulting PaO_2 of 289. Their calculated ideal PaO_2 would be approximately 300 (FiO_2 X 5) so this tells us oxygenation is fine, the reason for the hyperventilation is just that.... ventilatory, (removal of CO_2). The tidal volume should be based on ideal body weight, not actual weight in most instances. The fact the patient has become obese didn't change the size of their thoracic cavity and lungs. A 5'4" female that's 32 years old should weigh approximately 120lbs or 54kgs. With a standard tidal volume of 6-10ml/kg this gives us a range of 324ml-540ml. Starting at 500ml is reasonable. PEEP is not indicated in this patient thus far. The PaO_2 does not suggest inadequate oxygenation requiring increased mean airway pressures and nothing about the diagnosis triggers an automatic need for PEEP. Again starting with 0 PEEP is reasonable.

If you're having trouble with this question just consider my primary goal writing it was to emphasize do not paralyze the respiratory compensation patient and consider modes other than SIMV in your ventilatory management that will permit adequate compensation. When you see exam questions with answers you don't agree with, see if you can identify the information they're trying to verify your knowledge of. That might help you select the answer better. Considering the dynamic nature of medicine, many times test takers are more educated or well read than the item writer on that specific topic was at the time the question was designed. You may be able to anticipate the question's answer because you know the industry has taught "XYZ" regarding this issue for years. You also happen to know that the latest research, good research, research you were directly privy to, has debunked "XYZ" and you now know that "UVW" is a better answer selection. Don't answer "UVW" with the attitude, "I know they want "XYZ" but they're wrong and I'm standing my ground, "UVW" is right!" I'm sorry to inform you but the answer will still be keyed "XYZ" and you'll still get it wrong. Send your complaint at a later time, don't sink yourself on the exam because you happen to have late breaking, cutting edge information that the item writer did not have at the time of exam design.

4. A Both of these patient types will be extremely dehydrated and fluid resuscitation is critical. We don't 'rapidly' decrease glucose levels in anyone. Limit your decrease to 100mg/dL/hour, especially in pediatric patients. Per the last question, paralysis of these patients is typically not needed and potentially very harmful. Correction of acidosis is key in the DKA patient, not quite so in the HHNC patient. Remember the HHNC patient does not enter ketoacidosis.

5. D See question for rationale.

6. D I tried to 'sucker' you into answering one of the others by placing 'diabetes insipidus' with a group of questions about diabetic ketoacidosis. Remember the two pathologies are completely different. Diabetes insipidus (DI) is due to inadequate or no antidiuretic hormone (ADH) production/release from the pituitary. Commonly this results from head injuries and some medications like phenytoin (*Dilantin*®).

TEST TIP
diabetes insipidus (DI) = inadequate ADH (vasopressin)
DI caused by head injuries and Dilantin

7. B See rationale previous question.

8. D Because these patients are losing enormous amounts of water the urine becomes very hypo-osmolar ("watered down").

9. D Central neurogenic DI is most common followed statistically by nephrogenic DI.

10. C SIADH is an overproduction/release of ADH from the pituitary or from specific carcinomas that produce ADH like substances. As a result the ADH stimulates the aldosterone receptors of the nephrons resulting in increased sodium retention with secondary water retention. The result is a dilutional hyponatremia.

TEST TIP
syndrome of inappropriate antidiuretic hormone (SIADH) = too much ADH (vasopressin)
SIADH caused by head injuries & oat cell (small cell) carcinoma

11. C Small cell (oat cell) carcinoma is notorious for a couple things. First metastasis to the brain is very quick and presents with complications such as seizures and altered mentation. Second, oat cell carcinomas are very good at producing chemicals that resemble our own hormones, especially vasopressin and oxytocin. Lastly you should consider that anyone with cancer is hypercoagulopathic. Cancer equals clots. So look for embolisms in anyone with this diagnosis. In fact the patient that presents with a pulmonary embolism without any other obvious risk factors is assumed to have an undiagnosed cancer until proven otherwise.

TEST TIP
pregnancy, cancer, bedridden & surgical patients, females that smoke and use birth control
⇓
hypercoagulopathic ⇒ pulmonary embolisms

12. B Remember with SIADH they are releasing too much ADH (vasopressin) and thus holding sodium and water. More IV fluid and vasopressin are the last things this patient needs. An "aldosterone substitute" would not help either since this would again encourage reabsorption of water. These patients need diuresis and fluid restriction.

13. D The OG/NG tube option may seem a poor idea due to the likelihood of esophageal varices and possibility of trauma to these with insertion. This is a reasonable and commonly questioned issue. You must consider however that a GI bleed is depositing blood into the stomach. What will this cause? Right.... vomiting. Vomiting will result in extreme esophageal pressures which are much more likely to rupture a varices and kill the patient ultimately. Careful OG/NG tube insertion with continued gastric evacuation is essential. Octreotide (*Sandostatin*®) will cause vasoconstriction of the upper GI system and decrease the motility of such. It also decreases the hepatic portal pressures allowing better vascular evacuation of the upper GI system, all of which contribute to decreased bleeding.

14. D Remember vasopressin is the most potent vasoconstrictor made by the body and acts systemically. If the patient has a CAD history their pump may be too fragile/sick to handle the increased SVR that will be caused by vasopressin administration. Desmopressin (*DDAVP®*) is frequently used as an alternative to vasopressin. It is a synthetically derived version of vasopressin which has significantly less vasoactive action versus it's anti-diuretic action. In fact it's most commonly prescribed in the U.S. for nocturnal enuresis (bedwetting). Desmopressin may also be used to treat hemophilia A and thrombocytopenia as it stimulates endothelial cells of the kidneys, promoting factor VIII and vonWillebrand factor (vWF) release as well as increased erythrocyte (RBC) production from erythropoietin release.

15. D Psychogenic hyperventilation is very possible except she's not hyperventilating. If you're an "old seasoned veteran" you may have some cynical predispositions towards this type of patient presentation. Don't let that influence your test taking abilities.

Anxiety during early pregnancy is an absolute consideration here except you have no information suggesting she's pregnant. Don't let the answer choices suggest findings that are not in the stem of the question. Again this is an item writer technique that preys upon your predisposed mentality towards this type of scene/situation and plays a trick on your minds processing of the actual question stem.

Cushing's syndrome is an excess of cortisol and results in a myriad of symptoms that are not typical of the case presented.

Grave's disease or hyperthyroidism is classically seen in young otherwise healthy females. Think of the thyroid as setting the 'idle speed' of your body. If it's overactive, so is the body. Temperature goes up, you use nutrients faster and thus lose weight, increasing the heart rate and respiratory rate, etc. Look for sudden onset of atrial fibrillation or other tachyarrhythmias in young females as a 'hint' for hyperthyroidism.

TEST TIP
Young female, palpitations, tachycardia/A-fib ⇒ hyperthyroidism

16. B The patient has been taking a substantial dose of steroids for a prolonged period of time. As a result her body's natural steroid production has dropped to essentially zero. She has placed herself into Cushing's syndrome, "iatrogenic Cushing's syndrome" to be exact. Now with sudden withdrawal of the steroids her body needs to function, she does not have the chemical makeup necessary to release glucose properly or respond to alpha and beta stimulation normally.

17. A See rationale for question 15.

18. D This patient has placed himself into an iatrogenic Cushing's syndrome. His excess steroid ingestion has caused the classic picture of such. You should anticipate obese habitus primarily in the truncal region, thin extremities as muscle wasting occurs there first. Bruising and abdominal striae are common along with the classic 'buffalo hump', which is most readily identified on females. Weakness and fatigue accompany hyperglycemia. Males will also complain of decreased libido.

19. C Acute pancreatitis is frequently accompanied by a history of alcoholism, sepsis and ARDS. Hopefully as you read the question stem you noted the respiratory distress, lab values (elevated bands) and chest X-ray findings to suggest sepsis and ARDS. Add in the abdominal pain with elevation of the left diaphragm and you've got a pretty good indication the source is the pancreas. If this didn't work for you, can you rule out the other options? Appendicitis is lower abdominal so that should be pretty easy. GI hemorrhage was a possibility with the history and nausea/vomiting but doesn't make sense with the respiratory changes. Hepatitis makes sense with the history and abdominal pain but again doesn't explain the respiratory changes and doesn't match the left diaphragmatic elevation.

TEST TIP
Cullen's sign, alcoholism, LUQ pain ⇒ pancreatitis ⇒ sepsis & ARDS

20. D The 'coarse muscle flapping' is a telltale for hepatic encephalopathy. Commit it to memory for certification exams. A and C are both reasonable suspicions but inadequately supported on the very brief question stem.

TEST TIP
"coarse muscle flapping" ⇒ hepatic encephalopathy

21. B The spleen is the most common abdominal organ injured in blunt trauma and must be treated aggressively. Kehr's sign refers to the abdominal pain that is referred typically to the left shoulder in this case.

TEST TIP
Kehr's sign ⇒spleen
Murphy's sign ⇒ gallbladder

22. C Murphy's sign is elicited by pressing up into the right upward quadrant trapping the gallbladder between the hand and the liver. The patient is then instructed to take a deep breath and if the breath is halted prematurely due to pain, the sign is positive and suggestive of gallbladder pathology.

23. A Brudzinski's sign is elicited by flexing the chin towards the chest. This tightening of the irritated meninges causes a reflex flexion or lifting of the legs in an effort to shorten the meninges and relieve the discomfort. This is similar to Kernig's in that the leg, flexed at the hip and knee, is straightened at the knee. If while straightening the knee, pain is felt in the neck or the patient flexes the neck in an effort to shorten the meninges, meningism is likely present.

24. B Levine's sign is the class 'clenched fist' over the chest sign you've seen in CPR classes for decades.

25. B See rationale question 23.

NEUROLOGICAL EMERGENCIES

1. The three meninges of the central nervous system include the;
   a. pia mater, dura mater, arachnoid space
   b. dura mater, arachnoid membrane, tentorial membrane
   c. tentorium, pia mater and dura mater
   d. pia mater, dura mater and arachnoid membrane

2. Cerebrospinal fluid (CSF) will be found;
   a. in the subdural space
   b. in the subarachnoid space
   c. in the epidural space
   d. all of the above

3. Cerebrospinal fluid;
   a. is synthesized during the fetal stage and first two years of life, then remains unchanged for the remainder of life
   b. is synthesized through adulthood or until the ischial plates stop growing, then remains unchanged for the remainder of life
   c. synthesizes and is reabsorbed every 24 to 72 hours
   d. synthesizes and is reabsorbed approximately three times per day

4. Cerebrospinal fluid
   a. is hypertonic
   b. is hypotonic
   c. is isotonic
   d. none of the above

5. The tentorium is;
   a. part of the dura mater and forms a dividing membrane between the upper and lower brain
   b. part of the arachnoid membrane and forms the CSF reabsorbing mechanism
   c. the outermost lining of the pia mater and is very delicate
   d. the outermost lining of the dura mater and is adhered to the skull

6. Normal ICP is _____ and normal CCP is _____.
    a. 0-20mm Hg, less than 60mm Hg
    b. 0-10mm Hg, 70-90mm Hg
    c. 10-15mm Hg, greater than 60mm Hg
    d. 0-10mm Hg, greater than 60mm Hg

7. Your patient has the following hemodynamic parameters. HR 64, BP 180/90, RR 22 and irregular. ICP 21, CVP 20, PAP 32/16, PCWP 16. (All pressures are in mm Hg). Your patient's CPP is;
    a. 69mm Hg
    b. 79mm Hg
    c. 89mm Hg
    d. 99mm Hg

8. Bonus question* Based on the information provided in the previous question, what is the coronary perfusion pressure? Remember the cardiac section?
    a 64mm Hg
    b. 74mm Hg
    c. 90mm Hg
    d. 160mm Hg

9. You could calculate a mean arterial pressure (MAP) using the formula;
    a. MAP=(SBP + DBP)/3
    b. MAP=((2SBP)-DBP)/3
    c. MAP=2(SBP-DBP)/3
    d. MAP=DBP + 1/3(pulse pressure)

10. Signs of acute neurological insult include all but which of the following;
    a. decorticate posturing
    b. decerebrate posturing
    c. defensive posturing
    d. ipsilateral pupillary dilation

11. Cushing's Triad consists of;
    a. widening pulse pressure, tachycardia and respiratory changes
    b. hypertension, bradycardia and widening pulse pressures
    c. hypertension, tachycardia and respiratory changes
    d. hypertension, bradycardia and respiratory changes

12. All of the following would be appropriate management options for the head injury patient except;
    a. Provide neutral head alignment with 15-30 degree reverse trendelenburg positioning
    b. Provide adequate neuromuscular blockade prior to placing an OG tube or foley
    c. Provide protection from noxious stimuli by using ear muffs/plugs, blindfolds and adequate padding
    d. Coordinate with the pilot reference climb rate, cabin pressurization and takeoff/landing procedures to limit ICP spikes.

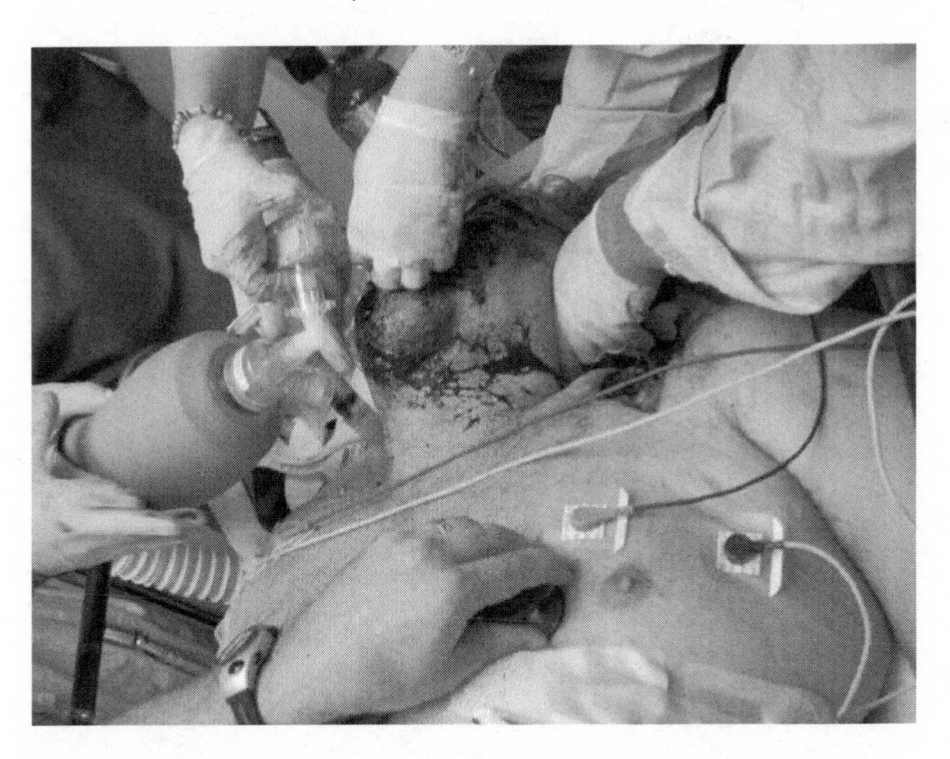

13. Your patient received a closed head injury when he was shot with a pistol during a hunting trip. The initial CT report from the sending facility states that he has a single foreign body located in the right occipital lobe of the brain with a leftward midline shift of 6mm. He has an intracerebral bleed with collapse of the right lateral ventricle, bleeding appears to be stabilized at this time. An epidural "Camino" (ICP monitor) has been placed and transduced readings indicate an ICP of 44mm Hg. Vital signs are HR 54, BP 170/50, RR 14 (controlled via ventilator), SpO_2 99% (with FiO_2 0.6), $EtCO_2$ 31. (ABG's drawn on your arrival show; pH 7.39, $PaCO_2$ 42, PaO_2 288, HCO_3 26). Which of the following would be the best management plan.
    a. reduce ICP and increase CPP by venting the epidural Camino to an ICP of 10-12mm Hg while monitoring BP and HR for sudden changes. Stop venting if hemodynamic instability ensues.
    b. reduce ICP and increase CPP by administering a vasodilator titrating to a SBP less than 160 but do not drop the pressure any lower.
    c. reduce ICP and increase CPP by hyperventilating the patient to an $EtCO_2$ of 25 and restrict fluids. Administer antihypertensive agents if SBP exceeds 180 or DBP exceeds 90.
    d. reduce ICP and increase CPP by hyperventilating the patient to an $EtCO_2$ of 24 and administer fluids and/or vasopressors as needed to keep the BP at or above it's current values.

14. A subdural hematoma is typically ______ in nature and lies between the _________.
    a. arterial ; dura mater and pia mater
    b. venous ; pia mater and arachnoid membrane
    c. venous ; epidural space and the dura mater
    d. venous ; dura mater and arachnoid membrane

15. Subacute subdural hematomas become symptomatic;
    a. almost immediately
    b. in 1 to 2 days
    c. in 2 to 10 days
    d. in 1 to 6 months

16. The chronic subdural will commonly demonstrate a(n) ________ pattern on CT.
    a. acute hyperdensity
    b. diffuse patchy infiltrate
    c. generalized hypodensity
    d. granulated or 'salt and pepper'

17. Diffuse axonal injury (DAI) will commonly be ________ on CT.
    a. granulated or show a 'salt and pepper' appearance
    b. demonstrate a generalized hypodensity pattern
    c. unidentifiable from normal
    d. identified by cervical spine subluxation

18. Patients most likely to demonstrate an acute subdural include;
    a. pediatric patients
    b. adult patients
    c. elderly patients
    d. all patients are equally likely to demonstrate an acute subdural

19. Signs and symptoms of the subdural hematoma in a pediatric patient might include all of the following except;
    a. bulging sclera
    b. bulging fontanelle
    c. retinal hemorrhage
    d. unresolved lethargy

20. Epidural bleeds are not;
    a. more commonly arterial than venous
    b. more acute than chronic in nature
    c. typically involved with a lucid interval
    d. typically caused by a medial meningeal vein rupture

21. Your patient, a 76-year-old male who lives alone fell two days ago striking his head against the floor. A family member assisted him up. He was observed for the next 12 hours and did not show any further changes nor did he voice any complaints. He was visited regularly over the next three days with slight changes in behavior noted the third day. Now, day four, he presents with obviously altered LOC, slow/regular respiratory pattern, HR 55 and a BP of 168/78 with a blood glucose level of 68. His prior history includes multiple hip replacements, alcoholism and mild COPD. Based on this information, he would be most likely diagnosed with which of the following neurological problems?
    a. classic concussion
    b. chronic subdural
    c. subacute subdural
    d. diffuse axonal injury

22. Your patient was admitted to the ED with a chief complaint of severe headache, lethargy and reports of a stiff neck. She states moving her head makes the headache worse. The physician has determined the two most likely diagnoses are subarachnoid hemorrhage or meningitis. Which of the following tests should be performed next to best identify the etiology?
    a. lumbar puncture
    b. computed tomography (CT) scan
    c. CBC with differential
    d. blood culture

23. The second most common cause of a subarachnoid bleed is a 'berri aneurysm' rupture. The most common cause is;
    a. hemorrhagic stroke
    b. meningeal artery rupture
    c. uncal herniation
    d. trauma

24. Your patient's CT demonstrates hemorrhage in the white matter of the frontal region of the brain. He experienced traumatic shearing forces during a coup contrecoup type injury pattern. This describes a(n);
    a. epidural hemorrhage
    b. subdural hemorrhage
    c. subarachnoid hemorrhage
    d. intracerebral hemorrhage

25. Your patient experienced an accidental strike to the head from a child swinging a baseball bat in the house. He fell to the ground dazed for a few minutes but did not lose consciousness. He currently has no recall of the actual strike but does remember handing the child the bat and telling them, "don't swing this in the house". He has experienced a;
    a. classic concussion
    b. mild concussion
    c. fatherly moment
    d. lucid interval

26. Management of a closed head injury is commonly referred to as "Triple-H therapy". This consists of;
    a. hypertension, hypervolemia, hemodilution
    b. hypotension, hypercalcemia, hemoconcentration
    c. hypertension, hyperoxygenation, hypocapnia
    d. hypotension, hyperoxygenation, hypercapnia

27. Three classifications of cerebrovascular accident (CVA) are;
    a. embolic, hemorrhagic, diffuse
    b. thrombotic, embolic, focal
    c. hemorrhagic, embolic, hypertensive
    d. embolic, hemorrhagic, thrombotic

28. With some CVA's, ________ therapy is indicated within ________.
    a. supportive , 1 day of onset
    b. blood replacement, 10 minutes of onset
    c. thrombolytic, 3 hours of onset
    d. embolic, 3 hours of onset

29. Your patient was struck with an unknown object during an assault. He experienced a loss of consciousness at that time and has subsequently been intubated. A notable skull depression found on palpation prompted a skull X-ray upon arrival which demonstrates a fracture focal point which has multiple fractures radiating outward from the focal point. This would describe which type of skull fracture?
    a. linear
    b. linear stellate
    c. diastatic
    d. eggshell

30. Your patient was involved in a motor vehicle accident in which he was not restrained and struck the dashboard with his face and head. He presents at the receiving facility with epistaxis, periorbital eccyhmosis with an altered level of consciousness and trismus. His vitals are BP 170/90, HR 110, RR 30 and regular, SpO_2 96% on non-rebreather. Further assessment finds that maxillary and nasal regions appear to be soft and moveable while the remainder of the face is apparently stable. Which of the following would be the most hazardous therapy?
    a. nasotracheal intubation
    b. nasogastric tube insertion
    c. mask ventilation assistance
    d. nasal cannula application

31. The patient in the previous question likely has;
    a. LeFort I & II fractures
    b. LeFort II & III fractures
    c. LeFort I, II & III fractures
    d. LeFort III fracture

32. Which of the following skull fracture complications is most significant to the aeromedical transport provider?
    a. infection
    b. hematoma
    c. cerebral tissue damage
    d. pneumocephalus

33. Your 18-year-old patient was involved in a motorcycle vs. car crash. He was helmeted and appropriately dressed. Unfortunately this didn't keep him from being ejected 45 feet coming to an abrupt halt when he hit a building head first. He presents with a slightly altered LOC, obvious right radial fracture and an open tib-fib fracture of the right leg. He denies pain to his injuries and repeatedly asks, "How's my bike?". His vital signs indicate BP 72/33, HR 54, RR 26, SpO_2 92%. From the information provided, the best initial diagnosis would be;
    a. acute closed head injury
    b. neurogenic shock
    c. drug/alcohol toxicity with limb fractures
    d. hemorrhagic shock

34. The best management of the patient in the above scenario would likely include;
    a. hyperoxygenation, fluid restriction, PASG application, spinal movement restriction
    b. hyperoxygenation, steroid administration, fluid administration, spinal movement restriction
    c. hyperoxygenation, fluid administration, spinal movement restriction, vasopressor administration
    d. hyperoxygenation, hyperventilation, steroid administration, vasopressor administration

35. Your patient has received a spinal cord injury and has no sensation below the xiphoid process. His injury is likely at or below;
    a. T4
    b. T6
    c. T8
    d. T10

KEY & RATIONALE

NEUROLOGICAL EMERGENCIES

1. D Basic anatomy, refresh with your favorite anatomy text if this wasn't a "gimme".

2. B D is an attractive option if you haven't refreshed on this lately. Remember that the subdural and epidural spaces are 'potential spaces' only.

3. D Cerebrospinal fluid is produced by the ependymal cells of the choroid plexus and circulates from the lateral ventricles to the third and fourth ventricles finally down the spinal cord only to return up to the brain and be reabsorbed at the ventral sinuses. Failure to reabsorb CSF at the same rate as it is produced is the mechanism behind many disorders such as hydrocephalus. This cycle occurs approximately three times per day in "normal man".

4. C If it was A or B, it would pull fluid or third-space, both of which would cause dangerous pathologies. CSF is almost identical to blood plasma, it can even be tested for blood glucose level.

5. A The tentorium forms a shelf like structure (tentorium or "tent like structure" in latin derivation) that separates the upper brain from the lower (horizontal membrane visible between the '2' and '1' in the picture at right). It is actually an inner lining of the dura mater with the outer lining adhered to the skull and extending down the spinal cord to the sacral regions. The oval opening in the middle of the tentorium (not clearly visible in the picture to the right), which allows the upper brain (2) to communicate through to the lower brain (1) is called the 'tentorial inscisura' or 'inscisura tentorii'. The part of the brain communicating through the inscisura is the 'uncus'. The term 'uncal herniation' refers to herniation of the upper brain through the tentorial inscisura into the lower

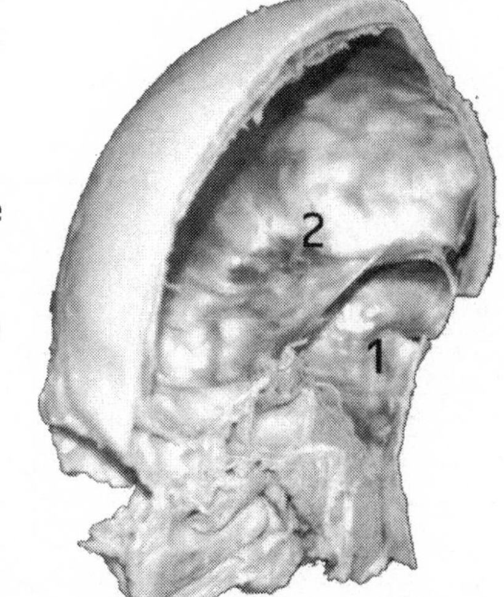

brain compartment and is the most common form of herniation. When a 'mass effect' occurs in the upper tentorial compartment (2) causing a shift of the brain downward into the lower brain compartment (1) this is called a 'descending herniation'.

6. B These are numbers used by the texts referenced for the exams we are focusing on. Other references may provide different values. You should know that normal CPP is 70-90mm Hg but we only target a CPP greater than 60mm Hg when treating a head injury patient.

TEST TIP
Normal ICP is 0-10mmHg
CPP = MAP - ICP
Normal CPP is 70 to 90 but we only need 60 for a head injury victim

7. D The first step in figuring out the answer is identifying how to calculate a CPP. CPP = MAP - ICP. This is a formula you must dedicate to memory. The second step is plugging in the values but you must calculate a MAP first. There are many formula permutations for MAP. Use any you like;

I used MAP = (SBP + (2 X DBP)/3
MAP = (180 + (2 X (90))/3
MAP = (180 + 180)/3
MAP = 360/3
MAP = 120

Next plug in the known values for CPP = MAP - ICP, or;
CPP = 120 - 21
CPP = 99

8. B Recall from the cardiac section that the coronary perfusion pressure is typically calculated as the diastolic less the wedge pressure or;

CoPP = AoDP - PCWP = 90 - 16 = 74mm Hg

9. D The only valid formula offered is D. This formula utilizes perfusion pressure, which is SBP - DBP. While this formula is not as commonly used (in my opinion), it is a valid formula and just happens to be the formula used in the exam reference texts.

TEST TIP
MAP = DBP + 1/3(pulse pressure)

10. C Defensive posturing is the posture I take when my wife asks me questions about our checking account. If you are not familiar with decorticate or decerebrate posturing you should review these terms and their implications in your favorite physiology text.

11. D D is the most commonly used definition however some texts will cite option B. Know that the exam references use hypertension, bradycardia and respiratory changes as the definition for Cushing's Triad also incorrectly referred to as Cushing's reflex, not to be confused with Cushing's Syndrome. Confused? Who's confused?

12. B While paralysis is commonly used to prevent dangerous physiologic complications to some stimuli like intubation, ventilation and the like, the use of NMBA's for placement of an OG or foley would be inappropriate. Optimally, you would want to use sedatives and/or analgesics to prevent the distress caused by these procedures.

13. D This patient scenario is a typical elevated ICP patient with mass effect in the upper right hemisphere secondary to trauma. The epidural Camino is acceptable for measuring pressures however the limitations of the epidural placement include decreased measurement accuracy and inability to aspirate the ventricles. Thus option A is physically impossible with this type of Camino. An intraventricular (aka- 'ventric') placement would allow the therapy described and would be possibly useful in this situation.

Option B may sound like a good idea but it is not. Old school training many years ago may have suggested this but the hypertension seen in head injury patients is compensatory and therefore required for continued survival. In fact if you calculate the CPP for this patient you'll find that it is only 46mm Hg. Our goal is to keep that CPP greater than 60mm Hg so dropping the BP from where it is would be potentially disastrous.

Option C may sound good to some of you. For years we were taught to hyperventilate head injuries with the mindset being that 'blowing off CO_2' would promote vasoconstriction and thus limit ICP. Did this work? Absolutely! In fact it worked so well that we vasoconstricted the entire brain and prevented adequate blood delivery to uninjured parts of the cerebrum, resulting in a cascade of cellular changes called 'secondary brain injury'. This prompted a paradigm shift in head injury management and all of a sudden we were teaching to no longer hyperventilate head injuries. Typical of human nature we over-reacted.

The appropriate way to manage these patients is with *controlled* hyperventilation. Hyperventilation is a great tool for ICP management and is used every day to treat these patients during anesthesia and subsequently in the ICU, but with control. It should be used to alter the patient's acid-base status to the alkolotic end of the normal range. So we'd like to hyperventilate the CO_2 down until the pH reaches around 7.45 which will require the $PaCO_2$ to be around 34mm Hg (assuming there are perfect gases to begin with and nothing else is affecting acid/base arrangement... a fairy tale scenario I know. In reality this is usually guided by serial ABG's and ICP readings).

Option C would drop the $PaCO_2$ to around 36mm Hg, which would result in a pH change to approximately 7.44. (Refer back to the Acid-Base, Vent chapter for these calculations) Pretty good ventilation decision. The problem with option C is that keeping the BP below 180/90 is not necessarily a good idea, it's completely dependent upon the patient's ICP. At 180/90 the MAP is 120. If the ICP were greater than 60mm Hg the pressure would need to be higher to continue an adequate CPP. Granted an ICP of greater than 60mm Hg is pretty significant but it is not uncommon and you need to be able to follow the logic.

Finally, option D. Hyperventilating to an $EtCO_2$ of 24 will take the $PaCO_2$ to 35. This drop of 7mmHg in $PaCO_2$ will increase the pH approximately .056 taking it to ~7.45, the alkolotic end of the normal range. Since the current MAP is 90 and the ICP is 44 the CPP is 46. We don't want the BP to drop from its current state until the ICP is dropped; in fact we need the MAP to increase ~20 ideally. Keeping the head injury patient properly hydrated and adequately perfused is critical in TBI management. In reality, most physicians would have this patient hyperventilated into a slightly alkolotic state and the BP supported with pressors at or slightly above the current level in an attempt to meet the perfusion needs. Unfortunately, when patients get to this level of hypertension and ICP, there simply is no good non-surgical treatment and they have a poor prognosis.

14. D Subdural hematomas are most commonly venous in nature but they can be arterial. "Sub dural" indicates the hemorrhage is below the dura mater and above the arachnoid membrane exploiting the potential space.

15. C Subdural hematomas are categorized into acute, subacute and chronic. Acute are identified by becoming symptomatic with 24 hours, subacute are symptomatic in 2-10 days and chronic are typically symptomatic only after 2 weeks post injury.

16. D Chronic subdurals typically bleed for a prolonged period before becoming symptomatic, hence the 'chronic' categorization. During this time some of the blood is reabsorbed resulting in a 'salt and pepper' appearance on CT in the area of the brain experiencing the hemorrhage.

17. C Diffuse axonal injury (DAI) is the most severe category of concussion and frequently results in permanent major deficits. DAI is very difficult to identify on CT and as such CT is rarely used to confirm such a diagnosis.

18. A Refer to the categorization of subdurals in the rationale for question 15. Pediatric patients have a very 'tight head' meaning the entire calvarium is filled with meninges, CSF, blood and brain tissue. Any mass effect from hemorrhage will quickly cause changes in the brain tissue's perfusion thus becoming symptomatic faster. The elderly typically demonstrate more chronic subdurals secondary to the brain's atrophy, as we get older. This atrophy leaves more free space potential for 'extra luggage' like a hematoma.

19. A Bulging fontanelle(s) or retinal hemorrhages are classic signs of elevated ICP in the pediatric patient and associated with 'shaken baby syndrome'. Option D should have been obvious. Bulging sclera suggests prolonged elevated venous congestion.

20. D This question may be a little confusing due to the 'not' placed on the end of the stem. If the question stem is confusing you, try re-wording it in such a manner that it's easier to complete the phrase. The question is written to be read in such a manner that when the correct answer completes the stem it will make a true statement. So try reading the stem followed by each answer and decide at the end if you've made a true statement. Epidural bleeds are frequently associated with rupture of the medial meningeal <u>artery</u> in the temporal region. They are also typically arterial, not venous in nature; the 'lucid interval' is the period of lucidity following a loss of consciousness, which is then followed by a second loss of consciousness, and commonly death.

21. C This patient is showing symptoms of head injury four days after the actual incident. Based on that information alone and the diagnosis offered, subacute subdural is the best choice. The 'classic concussion' would require a history of some loss of consciousness at the time of incident. A chronic subdural would require the patient to remain asymptomatic for up to two weeks post injury. The diffuse axonal injury typically results in profound coma with profound long-term deficits.

22. B While the lumbar puncture is ideal to test for meningitis this would allow a 'venting' of the CSF fluid if it's under adequate pressure such as with a subarachnoid bleed. Therefore performing a lumbar puncture or 'spinal tap' on a subarachnoid bleed can allow the hemorrhage to be without any counter force or tamponade and bleed uncontrollably. We actually see a similar "venting" phenomenon when an epidural anesthetic is given to a pregnant female and the dura is accidentally penetrated allowing a CSF 'leak' into the epidural space. The result is a 'spinal headache' where the pressures in the head become deranged as a result of CSF not providing adequate counter-pressure. CBC with diff and blood cultures can also help identify meningitis but they are either not definitive or take too long when considering the alternative is a subarachnoid bleed.

23. D Self-explanatory. Meningeal artery rupture is associated with epidural bleeds, uncal herniation results from a supratentorial mass effect.

24. D The key here is that the bleeding is "in the white matter", thus in the tissue, not a potential or subarachnoid space. Shearing forces or projectiles typically cause intracerebral hemorrhages, they typically occur in the frontal and temporal regions and most commonly in white matter vs. grey.

25. B While C is very tempting; the mild concussion is typical of this scenario. The key to identifying this from 'classic concussion' is that there was no loss of consciousness, which by definition must occur for a 'classic concussion'. Both conditions will typically have a short duration retrograde amnesia, the duration of which can suggest the degree of concussion (i.e.- the more he can't remember, the more significant the whack to the head). Remember a 'lucid interval' is the period between a loss of consciousness and second loss of consciousness commonly seen with the epidural bleed.

TEST TIP
"lucid interval" ⇒ epidural bleed

26. A With head injuries we need to perfuse that brain. Keeping the blood pressure up (hypertension), the tank filled (hypervolemia) and the blood thin (hemodilution) will help perfuse everything the best. If you selected C, that was a very intuitive decision and makes perfect sense, it's just not what "Triple-H" therapy is referring to.

27. D Standard CVA definitions.

28. C Remember with embolic strokes we can administer thrombolytics to re-perfuse the brain. As of the writing of this text, the AHA guidelines recommend this not be performed outside of a three-hour window from time of onset of symptoms.

29. B The linear stellate fracture, an example seen below, typically has a focal depression with fractures radiating outwards, similar to a 'starred windshield' type pattern. Linear fractures, obvious to their name, are a simple fracture forming a single line across the skull surface. Diastatic fractures involve large sections of the skull being fractured from the main structure and becoming altered in their planar relationship to one another. Imagine a head crushed by a heavy object with the various skull fragments or plates in various planes of alignment. The term 'eggshell' fracture is commonly used to describe linear stellate and sometimes diastatic, it is not a technical fracture identification.

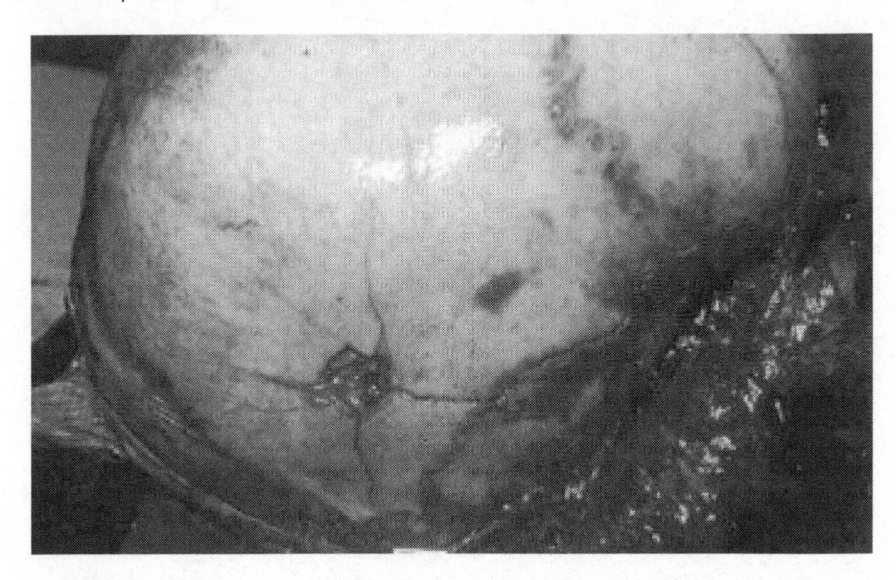

30. B The injuries should lead you to suspect basilar skull fracture with the apparent maxillo-facial fractures. Basilar skull fractures can involve the cribiform plate and result in allowing anything passed into the nasopharynx in a blind fashion to *possibly* enter the calvarium injuring the brain. The literature available has clearly demonstrated this is much more likely with an NG tube because of its lack of protrusion predictability and inability to assess the tip placement during actual insertion. Nasally inserted endotracheal tubes have been documented (almost exclusively in Europe), with similar complications but these were complicated by poor insertion technique and decision-making. Nasal intubation of the cranium with an endotracheal tube is very unlikely to occur if technically appropriate insertion technique is utilized.

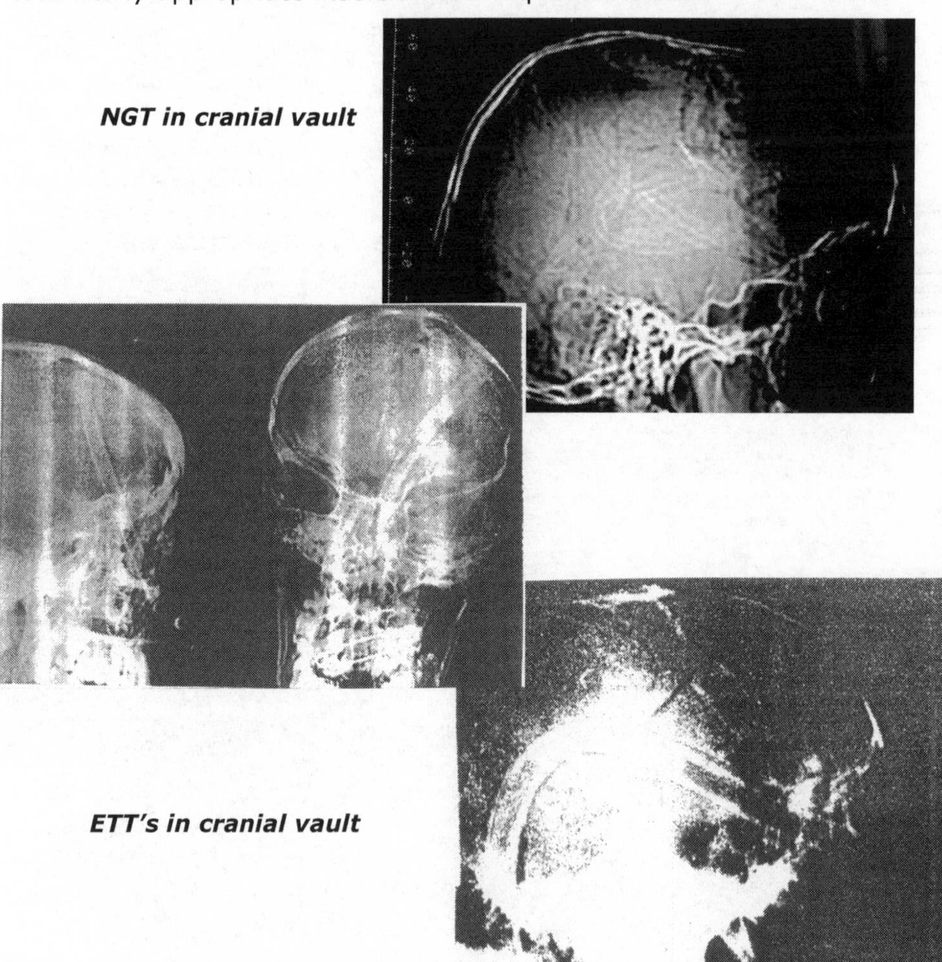

NGT in cranial vault

ETT's in cranial vault

31. A The LeFort classifications of fractures are as follows;

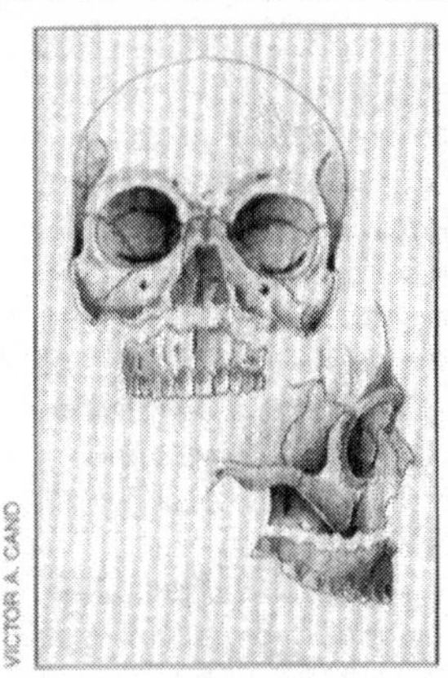

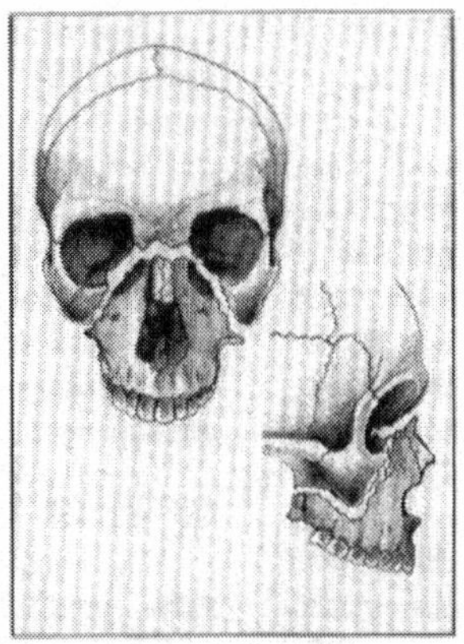
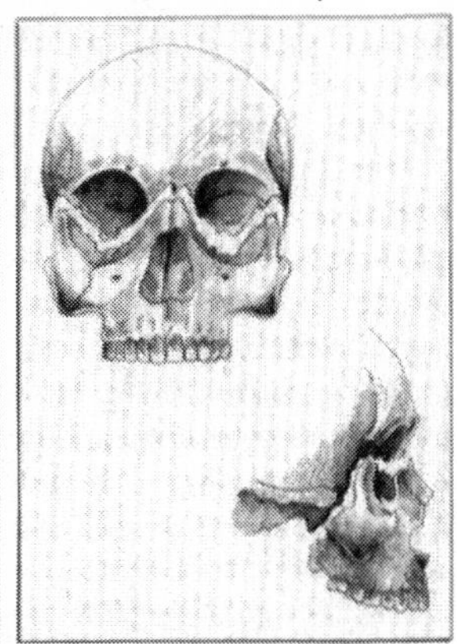

LeFort I ***LeFort II*** ***LeFort III***

The LeFort I results in a loose maxillary region or upper lip movement. The LeFort II or 'pyramidal' fracture results in a nasal section that is loose from the face itself and finally with a LeFort III, the entire face appears to be detached and mobile from the rest of the skull. All of these fractures cross various branches of the 5th cranial nerve, the trigeminal nerve, and the subsequent stimulation of that nerve results in trismus.

32. D While all of these issues are important to consider, the most important for the aeromedical transport provider must be the pneumocephalus. Your patient is at risk for acute ICP increases during climb-out as the cabin pressure drops. These patients should only be aeromedically transported in pressurized cabins when possible and careful communication with the pilot reference cabin pressure control is important.

33. B The key here is you have an otherwise healthy young male with a 'shocky' BP and <u>no</u> tachycardia. That should point you at neurogenic shock every time. Does the mechanism support the diagnosis? Yes! Could it be a head injury, absolutely but the vitals don't support an acute head injury. Could it be hemorrhagic shock, well it could but that leaves the lack of tachycardia unexplained. Remember that in the face of the hypotension the heart rate will increase *if it can*. In this case the heart is never receiving the message. If you answered C you're an old cynic like myself and probably tired of reading these questions. Take a break.

34. C We need to oxygenate, perfuse, fill the tank and limit further injury. Steroids were very popular for awhile and are still used in some places but their use is not supported by the latest research.

35. B A basic knowledge of the dermatomes will potentially help you on these exams. An easy way to remember these is;

Nipple line	T4
Xiphoid	T6
Umbilicus	T10
Pubis	T12

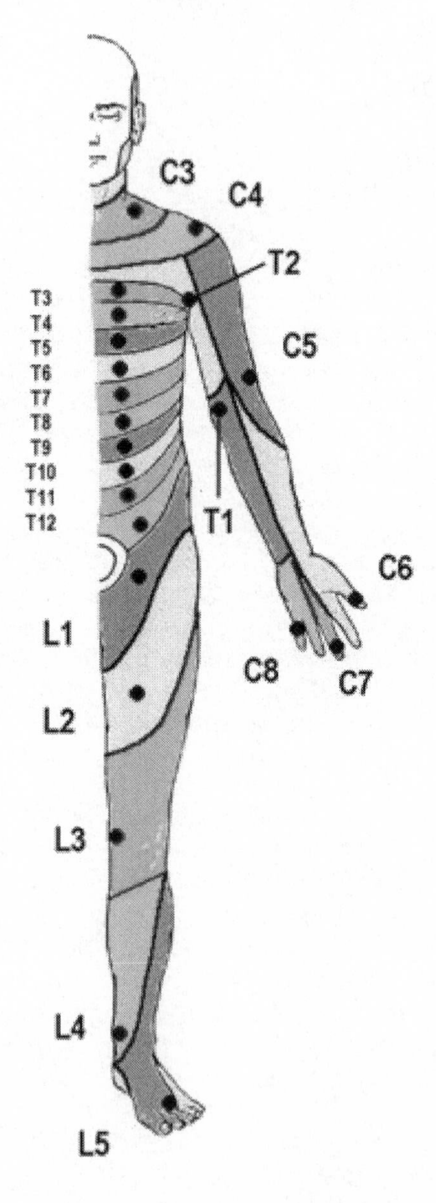

PEDIATRIC / NEONATAL EMERGENCIES

1. Which of the following techniques will not provide you with an appropriate endotracheal tube size for your 2-year-old patient?
   a. match ETT to the size of the nares
   b. (Age + 16)/4
   c. (Age/2) + 12
   d. match ETT to 5th digit diameter

2. Which of the following techniques will provide you with an appropriate endotracheal tube depth for your 2-year-old patient?
   a. match ETT to the distance from the earlobe to the corner of the mouth
   b. (Age + 16)/4
   c. (Age/2) + 12
   d. ETT diameter multiplied times two

3. Which of the following is not a true anatomical difference between pediatric and adult patients?
   a. pediatric lung tissue is more fragile than adult
   b. the mediastinum is more mobile in a pediatric patient vs. an adult
   c. the liver of a pediatric patient is proportionally larger than that of an adult
   d. the atria of the heart are larger than the ventricles in a pediatric heart vs. an adult

4. Your pediatric patient may have natural closure of the anterior fontanelle at _____ months of age and the posterior should close at _____ months of age.
   a. 10 ; 2
   b. 14 ; 18
   c. 12 ; 6
   d. 15 ; 2

5. Children have a larger volume of distribution and ________.
    a. larger tidal volumes
    b. lower hepatic flow rates
    c. higher cardiac output
    d. lower metabolic rate

6. Children have a relatively larger body surface area. This fact most significantly effects;
    a. sunburn/exposure likelihood
    b. temperature loss
    c. heart rate
    d. respiratory rate

7. With acute blood loss, the pediatric patient will not demonstrate hypotension until approximately ____ loss of circulating blood volume.
    a. 15%
    b. 25%
    c. 33%
    d. 45%

8. To improve cardiac output, the pediatric patient relies almost exclusively on;
    a. increased heart rate
    b. increased stroke volume
    c. increased preload
    d. increased vasoconstriction

9. Which of the following is an inaccurate statement regarding the pediatric airway?
    a. Peds typically have a relatively larger tongue in comparison to adults.
    b. The pediatric patient's higher larynx and shorter chin result in a more anterior relationship when visualizing to intubate.
    c. The vocal cords have a higher attachment point
    d. The narrowest part of the pediatric airway is the cricoid cartilage vs. the vocal cords in the adult patient.

10. Cuffed endotracheal tubes;
    a. Should never be used in pediatric patients
    b. May be used in pediatric patients but require deeper placement
    c. May only be used on patients over 5 years of age
    d. May be used in pediatric patients but should have a more distal cuff placement on the ETT for optimal safety.

11. The full-term neonate patient would most appropriately be intubated upon delivery with a size _____ endotracheal tube.
    a. 2.5-3.0
    b. 3.0-3.5
    c. 3.5-4.0
    d. 4.0-4.5

12. Which of the following diagnoses would not suggest difficulty with airway management?
    a. Pierre Robin syndrome
    b. Treacher-Collins syndrome
    c. Hurler's syndrome
    d. Duchenne's

13. Which of the following diagnoses would be a contraindication to the use of succinylcholine?
    a. Pierre Robin syndrome
    b. Treacher-Collins syndrome
    c. Hurler's syndrome
    d. Duchenne's

14. Standard fluid resuscitation for the infant or neonatal patient is;
    a. 10ml/kg, reassess and repeat PRN
    b. 20ml/kg, reassess and repeat PRN
    c. 40ml/kg, reassess and repeat PRN
    d. 50ml/kg, reassess and repeat PRN

15. You will be transporting a 2-year-old patient that weighs 12kgs. The appropriate IV fluid maintenance for this patient is;
    a. 25ml/hr
    b. 36ml/hr
    c. 44ml/hr
    d. 240ml/hr

16. Which of the following patients should be triaged for your management priority?
    a. 4 day old neonate, SpO_2 98% on blow by O_2, BGL 42
    b. 3 year old child, HR 138, temp 99.9°F, BGL 51
    c. 1 month old child, HR 190, temp 101.1°F, BGL 29
    d. 12 month old child, HR 72, SpO_2 98% on room air, sleeping quietly

17. Which of the following electrical therapies is most appropriate for a pediatric patient?
    a. Synch cardioversion at 2J/kg repeat at 4J/kg
    b. Defibrillate at 1j/kg repeat at 2J/kg
    c. Defibrillate at 2J/kg repeat at 6J/kg
    d. Synch cardioversion at 0.5-1.0J/kg

18. Which of the following resuscitation drug-dose combinations would be appropriate for a pediatric patient.
    a. atropine 0.01mg/kg, min 0.5mg, max 1.0mg
    b. epinephrine 0.01mg/kg, min 0.5mg, max 1mg
    c. atropine 0.02mg/kg, min 0.2mg, max 3.0mg
    d. lidocaine 1mg/kg, max 3mg/kg

19. A common injury pattern seen in children struck by a vehicle is called;
    a. Cushing's triad
    b. Winan's triad
    c. Wadell's triad
    d. Bernig's triad

20. The single largest cause of pediatric traumatic deaths is from;
    a. motor vehicle crashes
    b. drowning
    c. firearms
    d. child abuse/assault

21. Of all child abuse cases, the most commonly injured system is the ______, while the most common system injured causing death is the ______.
    a. neuro system ; integumentary system
    b. musculoskeletal system ; neuro system
    c. integumentary system ; neuro system
    d. genitourinary system ; neuro system

22. Which of the following fractures is typical of child abuse?
    a. Greenstick-Periosteum
    b. Comminuted
    c. Compressed
    d. Spiral

23. While transporting a patient in an isolette you note that the patient has become somewhat less responsive, or obtunded, bradycardic and slightly hypotensive. All oxygen delivery equipment is working properly and SpO_2 indicates 99%. Gentle stimulation does not seem to change the infants responsiveness or heart rate. Which of the following findings would be typical (or possibly causative) of this situation?
    a. blood glucose level 101mg/dl
    b. temperature 35.1°C
    c. excessive cabin pressure
    d. temperature 39.0°C

24. Reye's Syndrome has been associated with fever in children and which medication?
    a. acetaminophen
    b. acetylsalicylic acid
    c. ibuprofen
    d. diphenhydramine

25. You have decided to ventilate a four month old, 5.5kg female with status seizures. From the following, which vent settings for this patient would be best;
    a. Volume targeted ventilation, SIMV, Vt 60ml, F 30/min, PEEP 5cm, FiO_2 60%
    b. Pressure targeted ventilation, SIMV, PIP 20, F 30/min, PEEP 5cm, IT 0.5sec, FiO_2 100%
    c. Volume targeted ventilation, AC, Vt 44ml, F 20/min, PEEP 3cm, FiO_2 50%
    d. Pressure targeted ventilation, SIMV, PIP 10, F 30/min, PEEP 3cm, IT 0.1sec, FiO_2 80%

26. While transporting an infant to a tertiary care facility for treatment of a congenital heart defect you note that the patient appears to be 'acting funny'. You note a rhythmic 'bicycling' action of the legs with a fixed gaze up and to the left, which lasts approximately a minute. Your patient may have just experienced;
    a. a bowel movement
    b. a myoclonic seizure
    c. a subtle seizure
    d. a clonic seizure

27. You are transporting an infant from a rural facility to a tertiary NICU. Upon your initial examination you note diminished lung sounds in the left posterior basilar region, dyspnea and a slightly scaphoid abdomen. You would suspect;
    a. left pleural effusion
    b. left pneumothorax
    c. choanal atresia
    d. diaphragmatic hernia

28. Your patient with choanal atresia will have difficulty with all but which of the following;
    a. breathing
    b. feeding
    c. sleeping
    d. diuresis

29. You have just performed a field delivery of an infant that was previously calculated to be 34 weeks gestation. You note the newborn upon delivery has greenish-tinged mucous on the face, hair and upper body. As you begin to wrap towels around the newborn, he begins crying, has obvious movement of all extremities and his respiratory rate is approximately 40 and regular. Your partner auscultates a heart rate of 132 while you are doing this and remarks the baby's skin is pink at the core but slightly pale in the extremities. Your immediate course of action should be to;
    a. perform meconium aspiration suctioning with an endotracheal tube
    b. perform mouth then nasal suctioning using a bulb syringe
    c. perform nasal then mouth suctioning using a bulb syringe
    d. clamp and cut the umbilical cord

30. You are transporting a newborn with a preliminary diagnosis of trisomy-21. This 2-day-old male has had difficulty maintaining his oxygen saturation and developed 'wet lungs'. The patient has been intubated for transport, vital signs are currently within normal limits and appear to be stable. You would suspect this newborn is most likely to have;
    a. diaphragmatic hernia
    b. ventriculoseptal defect (VSD)
    c. pulmonary stenosis
    d. tricuspid insufficiency

31. The most common congenital heart defect is;
    a. patent ductus arteriosus (PDA)
    b. aortic stenosis
    c. Tetrology of Fallot (TOF)
    d. ventriculoseptal defect (VSD)

32. Congenital heart disorders of the 'cyanotic lesion' type would include all but which of the following;
    a. Tetrology of Fallot (TOF)
    b. transposition of the great vessels
    c. ventriculoseptal defect (VSD)
    d. truncus arteriosis

33. The primary stimulus which prompts closure of a PDA is;
    a. oxygen
    b. indomethacin
    c. prostaglandin
    d. carbon dioxide

34. A primary goal when transporting the patient with a cyanotic lesion is to;
    a. prevent hypercapnia
    b. hyperoxygenate
    c. limit opioid administration
    d. avoid noxious stimuli such as suctioning

35. Your patient was previously treated for his congenital heart defect receiving a Blalock Taussig Shunt (BT Shunt). This changes your therapy in that you;
    a. cannot initiate peripheral IV access in the upper extremities
    b. cannot use the effected arm for BP's and IV's
    c. cannot use typical tidal volumes during ventilation
    d. should use nitric oxide for any respiratory distress encountered

36. Which of the following would be an acceptable dosing regimen for Indomethacin when attempting to close a persistent PDA?
    a. 0.1mg/kg q 24hrs for three doses
    b. 0.2mg/kg with two more dosings at 0.1mg/kg
    c. 0.5mg/kg single dose
    d. 1mg/kg q 6hrs for three doses

37. Side effects to anticipate when administering prostaglandin (PGE_1) to maintain a PDA are all but which of the following;
    a. apnea
    b. bradycardia
    c. hypotension
    d. risk of delayed mental development

38. You are transporting a newborn diagnosed with a VSD. The patient reportedly has experienced clinically significant 'left to right shunting'. Which of the following clinical findings would you anticipate in this patient?
    a. jugular venous distention, clear lung sounds
    b. peripheral edema with cyanosis
    c. wet lung sounds and dyspnea
    d. hypertension and bradycardia

39. Your patient has been diagnosed with a coarctation of the aorta. During transport you would expect to;
    a. administer low concentration oxygen
    b. place the SpO_2 probe on the right hand only
    c. reverse the black and white ECG leads for a true Lead II view
    d. push IV fluids aggressively to maintain blood pressure within norms

40. Which of the following congenital disorders results in a 'right to left shunt'?
    a. Tetrology of Fallot (TOF)
    b. PDA
    c. isolated VSD
    d. transposition of the great vessels

KEY & RATIONALE

PEDIATRIC/NEONATAL EMERGENCIES

1. C A, B and D are all valid techniques for estimating endotracheal tube size. You should anticipate needing to calculate a pediatric tube size on the exams and be prepared to do so.

2. C This formula and the *appropriate** endotracheal tube size times *three*, will give you an acceptable estimate for tube depth. This formula tends to be slightly better for infants while the ETT size X 3 provides an easy, accurate estimate of tube depth for adult patients. Several studies have demonstrated that formula methods for determining pediatric ETT depth result in a malposition rate of 15-25%. Research has demonstrated that the highest rates for correct positioning result when the ETT is intentionally mainstem'd, then withdrawn until bilateral sounds are heard plus 0.5cm. It should be noted the research available was performed on 'normal' children without known congenital disorders, all of which were full-term gestational births.
*- identified by PALS recommended formula or by *Broselow®/Hinkle* tape

3. D A, B and C are all true along with the rib cage being more elastic and flexible and the skull bones are softer and separated by cartilage until approximately five years of age. Generally speaking bones are softer in pediatric patients and fractures should lead you to suspect higher energy mechanism of injury.

4. D The anterior fontanelle will typically close between 12 and 18 months. The posterior should close around 2 months of age.

5. C Pediatric patients should have smaller tidal volumes as their lungs are physically smaller than adult lungs. Their hepatic flow rates are actually the same or better than adults as they have not developed liver disease or CAD yet. Their cardiac output is generally higher with heart rate being the primary driving force behind this fact. Their metabolic rate is higher generally speaking which is my excuse for getting fatter as I get older.

6. B Temperature loss via radiation is a factor you must always consider when treating the pediatric patient. Any time a child is not responding to therapy, regardless of the therapy, ask yourself,

"Is the temperature right?". Hypothermia in the pediatric patient will complicate most acute pathologies in a significant fashion.

7. B Peds crash late but then they crash hard. It is the pediatric patient's resilience that keeps them from demonstrating measurable hypotension for a longer period but once this is exhausted, they rapidly decompensate. Heart rate is always your best indicator of compensation and adequate resuscitation of fluid losses.

8. A The pediatric patient has relatively little ability to change their contractility and thus stroke volume (SV). If you recall that cardiac output (CO) is stroke volume (SV) multiplied by heart rate (HR), CO = SV X HR. This might assist you in identifying what other factor will affect their CO.

9. C The vocal cords have a lower point of attachment in the pediatric airway. All other statements are accurate.

10. D Cuffed endotracheal tubes are acceptable in peds and are encouraged per the AAP and AHA now. Newer ETT's with more distal placement of the cuff (see right) are being introduced to allow for appropriate cuff placement without tube mainstem complications.

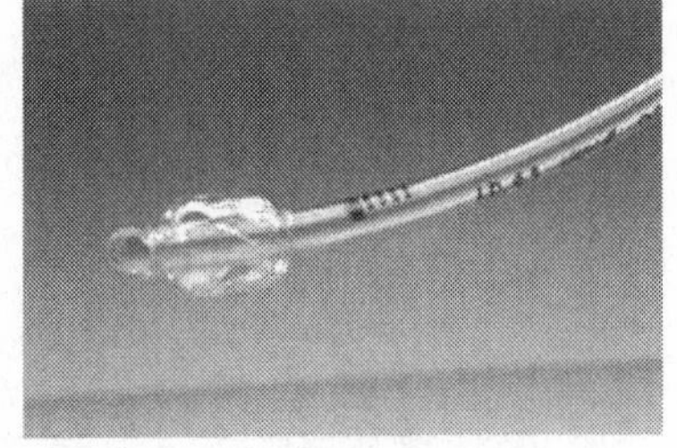

11. B

Premature neonates	2.5-3.0
Term neonates	3.0-3.5
3mos - 1yr	3.5-4.0
older than 1yr	use formula referenced in question 1

12. D Duchenne's muscular dystrophy is a form of muscular disorder and does not suggest airway complications per se. The other syndromes mentioned all involve actual airway deformities and alterations in anatomical relationship.

13. D Duchenne's muscular dystrophy is a form of muscular disorder in which proliferation of immature acetylcholine (ACh) receptors occurs. As a result, administration of SUX would promote a massive and commonly lethal release of potassium extracellularly. The other syndromes suggest difficult intubation therefore SUX would be a very reasonable selection for the RSI neuromuscular blockade.

14. A If you chose B, that would be accurate for peds over 1 year of age. Neonates and infants should receive 10ml/kg bolus with rapid reassessment.

TEST TIP
Fluid resuscitation >1 year old ⇒ 20ml/kg
<1 year old ⇒ 10ml/kg

15. C To calculate pediatric maintenance fluid rates you must use the 4/2/1 formula. First calculate the patient's weight in kilos. For the first 10kgs, administer 4ml/kg/hr. For the second 10kgs, administer an additional 2ml/kg/hr and for every kg over 20, administer an additional 1ml/kg/hr. So for example if your patient weighs 25kgs you would calculate their maintenance rate as follows;

4ml/kg/hr for first 10kg	=4ml X	10kg	=	40ml/hr
2ml/kg/hr for second 10kg	=2ml X	10kg	=	20ml/hr
1ml/kg/hr for every kg over 20	=1ml X	5kg	=	5ml/hr
TOTAL maintenance		25kg		65ml/hr

TEST TIP
Use the 4/2/1 formula for calculating pediatric maintenance rates.

16. C Recall that abnormal blood glucose levels (BGL) are;

neonate	less than 30mg/dl	(use D10 to correct)
child	less than 40mg/dl	(use D25 to correct)
adult	less than 70mg/dl	(use D50 to correct)

The way I remember this is the infant and child add up to the adult (30 + 40 = 70)

17. D Electrical therapy to be utilized per AHA PALS guidelines is;

- Synch cardioversion at 0.5-1.0J/kg and repeat at 2J/kg
- Defibrillate at 2J/kg, repeat at 4J/kg

To date, there has been insufficient evidence to support changes related to pediatric patients and cardioversion/defibrillation using biphasic cardioverter/defibrillator technology.

18. D For purposes of these exams you should know the standard pediatric resuscitation dosings for epinephrine, atropine and lidocaine. They are;

epinephrine	0.01mg/kg q3-5min
atropine	0.02mg/kg, min 0.1mg, max 0.5mg q3-5min
lidocaine	1.0mg/kg, q3-5min up to max 3mg/kg

TEST TIP
Pediatric medication dosages

Epinephrine	**0.01mg/kg 1:10,000 IV/IO**
-or-	**0.1mg/kg 1:1,000 ETT**
Atropine	**0.02mg/kg, min 0.1mg, max 0.5mg IV/IO**
Lidocaine	**1mg/kg , max 100mg IV/IO**
Naloxone	**0.1mg/kg, max 2mg IV/IO**
Prostaglandin (PGE_1)	**0.05-0.1mcg/kg/min IVgtt**
Etomidate	**0.2-0.4mg/kg, max 20mg IV/IO**

19. C Wadell's triad consists of injuries to the lower extremities as the child turns to face the vehicle in fear with the bumper striking the legs, the child then 'folds' over the hood causing chest and abdominal injuries, finally the child is then knocked away from the vehicle striking the head upon landing (kind of like a lawn dart).

20. A This is a long standing statistical fact. You may be asked this on the exams.

21. C The most commonly abused area is the skin while head injuries result in the most deaths.

22. D Spiral fractures in the arms are typical of twisting type forces and most common with child abuse related extremity fractures. Obviously skull fractures are very common with this patient population but none of the choices offered apply to skull fractures. You should review the various types of fractures seen in pediatrics to include; comminuted, compound, compressed, displaced, greenstick-periosteum, pathological, simple and spiral fractures.

23. B Remember our previous notes, when a child is not responding to therapy, think temperature. Obtunded, bradycardic, hypotensive kids are commonly cold. Fussy, irritable, tachypnic, tachycardic kids are hot.

24. B Although the research hasn't found a rock solid mechanism for the relationship between Reye's and aspirin, a statistically significant correlation has been found in multiple studies. Do not give aspirin to children with fever.

25. B Infants are typically pressure ventilated for safety purposes. Once older than one year of age, volume ventilation is common and acceptable. The safest overall mode of ventilation is SIMV as the patient with varying levels of sedation and/or paralysis best tolerates this. When pressure ventilating we must set a peak pressure and a PEEP. Commonly you'll see settings like 20/5 where 20 is the peak pressure the ventilator is to deliver before terminating the breath and 5 is the PEEP or minimum pressure the ventilator will allow the patient to exhale down to. Frequency (F) or rate should be physiologic normal and can be manipulated based on disease process. A four month old should be breathing somewhere around 24 to 40 breaths per minute. The inspiratory time (IT) is typically set at 0.5sec or greater to allow for adequate time to provide sufficient tidal volumes before the PIP reaches the pressure limit previously discussed/set. FiO_2 is typically set at 100% in infants requiring ventilation during transport. Infants that will be ventilated for a prolonged period of time and those that have previously been ventilated for a long time may be on a lower FiO_2 to prevent oxygen toxicity. For the relatively short duration transports seen in the transport setting, an FiO_2 of 100% will not typically cause oxygen toxicity. When you "pressure ventilate", Vt must be monitored closely to assure adequate Ve.

26. C Infants don't typically present with seizures the way adults and older children do. The classic tonic - clonic activities seen in the older patients is rather uncommon in infants because their neurological system hasn't developed well enough to support that type of presentation. Instead infants will commonly present with a *subtle* seizure, which commonly presents with repetitive mouth or tongue movements, bicycling of the lower extremities, eye deviations and repetitive blinking. Myoclonic seizures may involve multiple jerking motions typically in the upper extremities.

TEST TIP
"Subtle seizures" present with repetitive mouth/tongue movements, bicycling of the legs, eye deviations and repetitive blinking.

27. D The 'telltale' here is the 'scaphoid abdomen'. When you hear that term think diaphragmatic hernia first. The most common form of this disorder (a Bochdalek hernia - 95% occurrence rate) occurs at the left posterolateral region.

TEST TIP
"scaphoid abdomen" ⇒ diaphragmatic rupture/hernia

28. D Choanal atresia is a birth defect in which the posterior nasal passage is obstructed by tissue and sometimes cartilage or bone. Because infants are predominantly 'nose breathers' they will have difficulty with anything requiring them to breath via their nose. While this defect is easily managed with a simple oral airway or orotracheal intubation until surgically corrected, tasks such as feeding/suckling and sleeping are problematic. (The only other thing baby's do is poop so they basically find everything except testing diapers difficult.)

29. B This is an obviously meconium stained baby and all you 'old school' medics probably went straight for the meconium aspirator and suctioned. That was the teaching many years ago however current standards recommend only suctioning the infant with staining if they are not 'brisk'. The fact the child is crying, breathing adequately with a good heart rate and active muscle tone suggests that the child is brisk. Remember when you suction a newborn with your bulb syringe it's always mouth first, and then nose. They are obligate nose breathers and if you stimulate the nose first, that may prompt them to take their first breath inhaling the un-cleared oropharynx.

30. B Trisomy-21, (aka-Down's syndrome) is frequently associated with VSD's accompanied by aortic stenosis and PDA's. This is especially true in males.

31. D The ventriculoseptal defect is the most common form of congenital heart defect. Frequently this defect is without significant complication and effectively 'takes care of itself' with time. Auscultation of the heart should reveal an audible systolic murmur. The louder the murmur the smaller the VSD generally speaking. It is because of the small size of the VSD that increased turbulence is created as blood moves through it, hence the louder murmur. Quiet murmur, bigger VSD, bigger problem potential.

TEST TIP
The most common congenital heart defect is the VSD

32. C Congenital heart lesions are categorized into cyanotic lesions and acyanotic lesions. Cyanotic lesions are those in which blood being sent down the aorta has somehow bypassed the lungs and not been adequately oxygenated. Acyanotic lesions have abnormal blood flow but there is adequate mixing of oxygenated and unoxygenated blood prior to entering the systemic circulation.

Some standard cyanotic heart lesions you should be familiar with include;

***Tetrology of Fallot* (TOF)** (seen right) involves a combination of four defects. Many combinations exist but the classic combination includes a VSD, pulmonary stenosis, an overriding or right shifted aorta and right ventricular hypertrophy. The pulmonary stenosis makes it difficult to get blood to eject into the pulmonary artery and thus through the lungs. The VSD and overriding aorta allows the blood from the right ventricle to preferentially enter the aorta instead of the pulmonary artery and this results in unoxygenated blood entering the systemic circulation. The right ventricle becomes hypertrophied as it requires more and more strength to overcome the left ventricles pressures being transferred left to right via the VSD. Eventually a predominantly right to left shunt occurs.

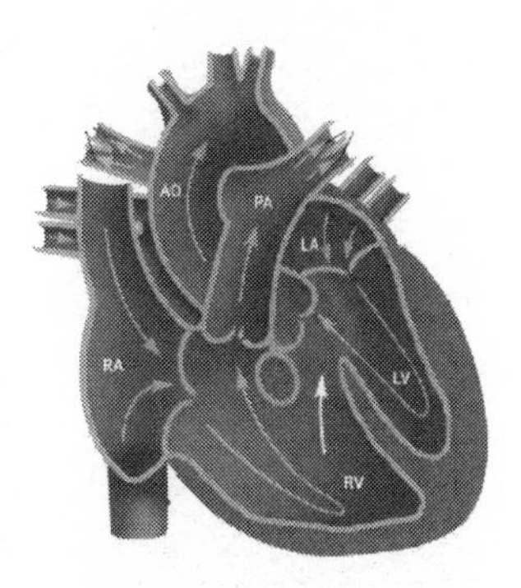

Transposition of the great vessels is just as it sounds (seen right). This condition is rapidly fatal unless there is a large PDA allowing communicating between the two isolated loops created by the transposition.

Truncus arteriosis (seen below) involves a defect in which the patient never developed both the pulmonary artery and aorta. Instead they formed a single large vessel that is shared between the pulmonary and systemic circulation. This is typically accompanied by a VSD with the single vessel sitting over the VSD allowing blood from the right and left ventricle to exit. Because the blood entering this single vessel is of significant pressure from the left ventricle, these high pressures tend to overwhelm the pulmonary circuit and create pulmonary edema with subsequent oxygenation issues.

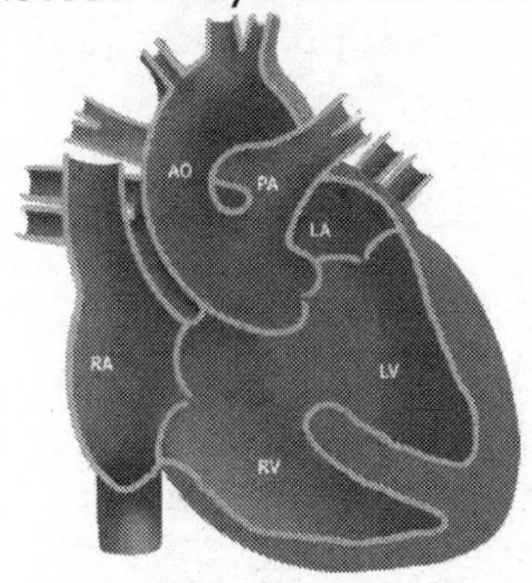

***Hypoplastic left heart syndrome* (HLHS)** (seen below left) is an extremely complex and difficult disorder to manage. This patient effectively has a single ventricle with the left heart poorly or not developed at all. The mitral and aortic valves are commonly absent, closed or very small and the aorta is also typically very small. Again these patients are extremely dependent upon a patent ductus arteriosus (PDA) as well as an atrial septal defect (ASD) allowing the blood returning from the lungs to cross from the underdeveloped left atrium back to the right atrium.

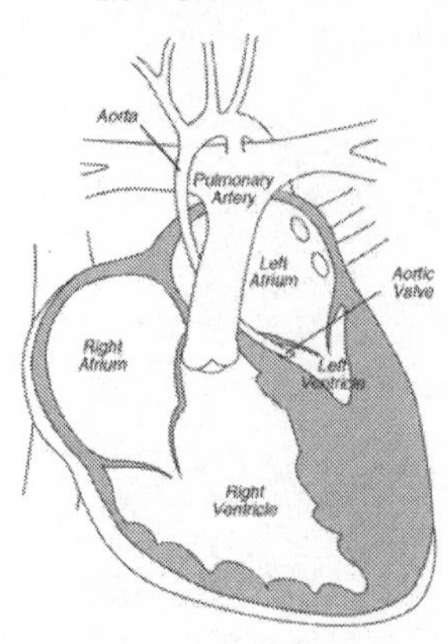

TEST TIP
Cyanotic lesions are typically PDA dependent

Some <u>acyanotic</u> lesions you should be familiar with include;

***Patent ductus arteriosus* (PDA)** (seen below left) is a failure of the PDA to close upon birth. This typically occurs within hours of birth but can take as much as a week. Patients with a symptomatic PDA typically develop symptoms around day three post delivery. A PDA alone is not a difficult disorder to manage and is not typically life threatening. The blood entering the pulmonary circuit from the aorta (down the pressure gradient) can begin to overwhelm the pulmonary system and results in CHF like symptoms. PDA's can be essential to survival in some defects (TOF, HLHS, etc.)

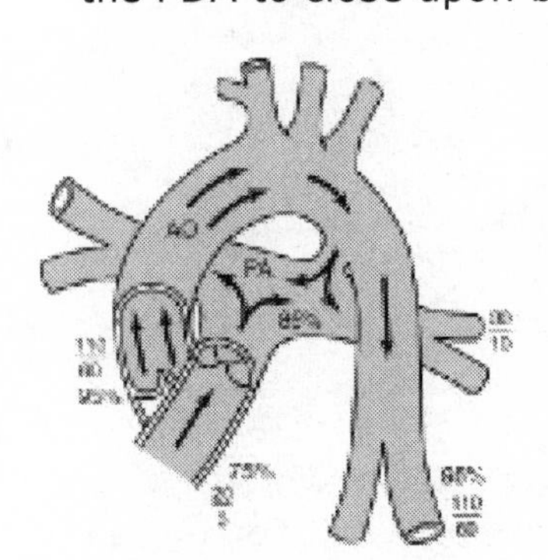

***Ventricular septal (aka-ventriculoseptal) defect* (VSD)** (seen left) is just as it sounds. The heart has a communication between the left and right ventricle. The size of the VSD will determine much of the symptomology and subsequent need for repair. Typically with an isolated VSD you will see a 'left to right shunt' meaning that the blood is moving from the higher pressure left ventricle (down the pressure gradient) to the lower pressure right ventricle. As a result, the right heart becomes overloaded leading to

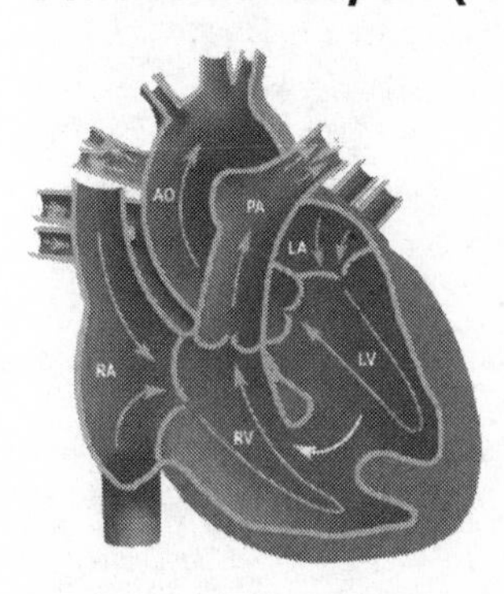

pulmonary congestion, CHF symptoms and right ventricular hypertrophy.

33. A Normally oxygen is the best 'stimulus' to prompt closure of the ductus arteriosus. As such, if the patient is "PDA dependent" because of other disorders, you may receive direction _not_ to administer high concentration oxygen. You may even receive orders to _not_ let the patient's SaO_2 exceed 90% or a similar number. The joke heard amongst NICU nurses is, "Oh my God his sat is 92%, quick, get a pillow!" Hey, don't get mad at me, I'm just repeating what I heard; I didn't make the joke up. (**Legal disclaimer: DO NOT smother the baby, under any circumstances.** I repeat, I am not advocating smothering a baby with a pillow. If you smother a baby with a pillow you're a sick freak and no, I will not come defend you in court.)

34. D You want to avoid any therapy that might promote pulmonary hypertension be it via coughing, straining, pain, acidosis, etc.

35. B A Blalock-Taussig shunt (BT shunt) by current day technique, involves placing a temporary shunt between the subclavian artery and the pulmonary artery (see right). This is typically used in cyanotic lesions such as tetrology of fallot. By placing this shunt, some of the lower saturated blood entering the systemic circulation is routed to the pulmonary circulation for oxygenation. Because this shunt 'robs' some of the affected limb of blood flow, that limb should not be used for procedures such as IV access and BP cuffs.

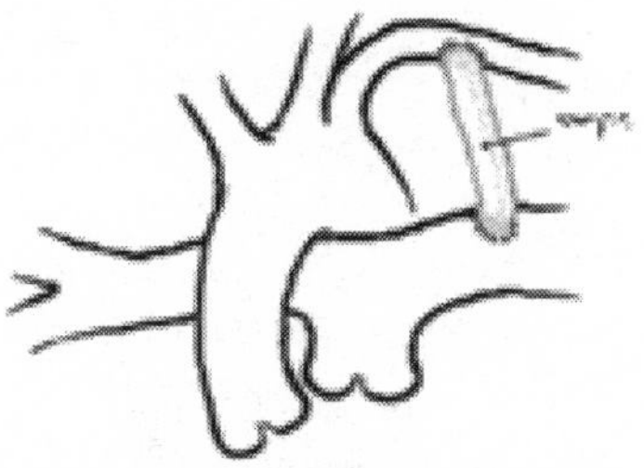

36. B Indomethacin dosing varies somewhat based on birth weight and time since birth. You will typically see one of two regimens used;

0.2mg/kg followed by 0.1mg/kg 12hrs after the first dose and repeated 24 hours after the second dose.

or;

0.2mg/kg q 12hrs for three doses

Remember that indomethacin is a prostaglandin synthesis inhibitor, which makes sense because prostaglandin is the drug of choice when we want to _maintain_ a PDA.

37. D Apnea, bradycardia and hypotension are all common side effects with the administration of prostaglandins (PGE_1).

38. C Recall that the left to right shunting via the VSD is causing an abnormally high flow and pressure into the pulmonary circulation, which subsequently causes CHF like symptoms.

39. B Because of the ***Coarctation of the aorta (COA)*** (arrow in diagram right), blood flow to the left subclavian may be significantly less than the right (assuming the coarctation occurs upstream of the left subclavian, not as pictured.) As such SpO_2 readings may be erroneous on the left hand. Ideally you would want all SpO_2 readings to come from the right subclavian's blood supply or the right hand.

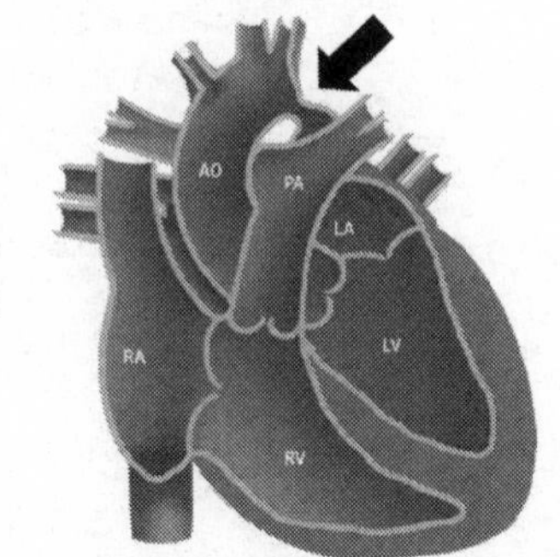

40. A Refer to the rationale and picture for question 32. With Tetrology of Fallot, the pulmonary stenosis, VSD and over-riding aorta promote blood movement from the right ventricle through the VSD and into the aorta, thus a 'right to left shunt'. PDA's and isolated VSD's will allow for a 'left to right' shunt. Transposition of the great vessels causes two isolated circuits that communicate hopefully via a PDA. Ideally in that case we want a 'left to right shunt' that allows the oxygenated blood on the left ventricular side to move retrograde via the PDA into the right ventricle's systemic circuit.

ENVIRONMENTAL EMERGENCIES, POISONINGS & TOXICOLOGY

1. The body's normal response to cold includes all of the following except;
   a. peripheral vasoconstriction
   b. skeletal muscle vasodilation
   c. increased cardiac output
   d. decreased metabolic rate

2. Shivering is limited by;
   a. thyroid hormone levels
   b. cardiac output
   c. muscle mass
   d. glycogen availability

3. Shivering will;
   a. increase cerebral metabolic oxygen requirements ($CRMO_2$) by 150%
   b. increase myocardial oxygen requirements (MRO_2) by up to 600%
   c. increase hemoglobin production by up to 200%
   d. increase myoglobin production by up to 200%

4. The ability to shiver is lost when core temperature drops below;
   a. 34°C
   b. 33°C
   c. 32°C
   d. 30°C

5. Your patient, a deer hunter, was found by search and rescue teams after being lost in the snow for two days. He has significant degrees of frostbite to his toes, fingers, nose and ears. His core temp when found was 29°C and his vitals were as follows, HR 49, BP 92/44, RR 10 and SpO_2 would not read. Which of the following lab values is most likely accurate for this patient?
   a. blood glucose level (BGL) 44mg/dl
   b. blood glucose level (BGL) 210mg/dl
   c. aPTT 14sec
   d. platelets 650,000

6. Which of the following lab/hemodynamic values would likely not be found in the patient described by the previous question?
   a. $PaCO_2$ 68mm Hg
   b. SVR 1680 dynes
   c. CI 1.1 L/min/m^2
   d. Bleeding time 1.5 minutes

7. ECG changes seen with severe hypothermia include all but which of the following;
   a. prolonged PRI
   b. prolonged QT
   c. Osborne waves
   d. delta waves

8. Severe hypothermia is designated when core temp falls below;
   a. 32°C
   b. 30°C
   c. 28°C
   d. 26°C

9. Level of consciousness will begin to fall when the patient enters;
   a. mild hypothermia
   b. moderate hypothermia
   c. severe hypothermia
   d. hypothermia does not affect level of consciousness

10. Your hypothermic patient is currently in cardiac arrest as verified by ECG. For airway management you should;
    a. perform gentle BLS maneuvers only
    b. insert only supraglottic airways such as a CombiTube, KingLT, PAX or PTL
    c. perform cautious/gentle intubation
    d. perform any maneuver other than intubation

11. In a hypothermic arrest scenario, you should hold medications until core temp reaches a minimum of;
    a. 32°C
    b. 30°C
    c. 28°C
    d. 26°C

12. The single most critical step in treating the hypothermia patient is;
    a. hyperoxygenation delivery
    b. careful and gentle handling
    c. consistent ventilation and chest compressions
    d. removal from the cold environment and appropriate re-warming

13. You have decided to begin active external re-warming of your patient. Which of the following is not a relevant concern here?
    a. chemical heat pack burns of the skin
    b. afterdrop
    c. sodium derangement
    d. rate of re-warming

14. Your hypothermia patient has been given room temperature IV fluid. You would anticipate the patient's temperature to _______ due to _________.
    a. rise ; convectional absorption
    b. drop ; conduction losses
    c. not change ; because of radiation losses
    d. improve ; conductional absorption

15. Your patient is reportedly an "undocumented alien" that was found by US Border Patrol walking in the June desert of southern New Mexico. He was found responsive, sweating profusely and weak. Per the Border Patrol agent's translation, the patient states he left Mexico two days ago and has been walking since, he adds he is 26 years old. The patient denies ever seeing a physician in his life and further denies any medical problems. He states he hasn't eaten in three days and ran out of water yesterday. As the patient begins speaking again, the agent shakes his head obviously in disbelief and says, "Now he's saying he can't breath and his chest hurts." The Border Patrol agents tell you this is the third time they've caught this subject crossing the border illegally in the last two months. You would anticipate and treat for which of the following first;
    a. profound dehydration
    b. heat stroke
    c. myocardial infarction
    d. heat exhaustion

16. As you begin treating the patient in the above scenario he begins to actively seize. The seizure is tonic-clonic and generalized in nature. It lasts approximately 3 minutes while you are aggressively attempting to oxygenate and obtain IV access. Because you're a 'high-speed' prehospital provider working for a cutting edge program, you have an I-Stat and quickly perform a blood draw and testing. Which of the following lab values would you anticipate finding in this scenario?
    a. Na 122
    b. BUN 8
    c. K 6.5
    d. CK-MM "LOW"

17. The seizure described in the patient of the prior two questions is especially problematic in this situation for all but which of the following reasons?
    a. increased MRO_2
    b. increasing temperature
    c. increased $CMRO_2$
    d. increasing glucose availability

18. Patients with profound hyperthermia can exhibit which of the following problems and it's related pathology?
    a. hypokalemia due to attempts to retain sodium
    b. acute tubular necrosis due to renal hypertension
    c. rhabdomylosis due to myoglobinuria
    d. elevated PT and aPTT due to overproduction of coagulation factors by the liver

19. Heat cramps are caused by;
    a. relative hypernatremia caused by water losses
    b. relative hyperkalemia caused by alkalosis
    c. relative hypoxia of muscle tissue resulting from inadequate perfusion
    d. relative hyponatremia resulting from ingesting only water

20. Heat exhaustion is defined as;
    a. an increase in core temperature without neurological impairment
    b. an increase in core temperature with neurological impairment
    c. any patient with an elevated core temp that has lost the ability to sweat
    d. any patient with an elevated core temp with neurological impairment that has the ability to sweat

21. Treatment of the heat stroke patient may likely include all but which of the following;
    a. aggressive cooling
    b. oxygen administration via simple mask
    c. aggressive fluid resuscitation with NS
    d. phenothiazines or NMBA's

22. Your patient was lost in the snow covered mountains for three days while searchers looked for him. He was found with a core temperature of 30°C. He was dry except for the lower pant legs which were wet with snow from walking. He would be classified as;
    a. an acute hypothermia victim
    b. a chronic hypothermia victim
    c. a dry hypothermia victim
    d. a wet hypothermia victim

23. Which of the following facts is inaccurate regarding heat stroke?
    a. core temperature exceeds 42°C
    b. level of consciousness is altered
    c. oxygen supply exceeds demand
    d. respiratory alkalosis occurs with metabolic acidosis

24. When ventilating the heat stroke patient;
    a. use of control mode is ideal with a rate not to exceed 20/min
    b. use of assist control mode is appropriate
    c. use of tidal volumes in the 15-20ml/kg range will assist with correcting respiratory alkalosis
    d. utilization of PEEP is necessary for optimal thermal energy transfer

25. Additional medications to consider when treating the heat stroke victim include;
    a. diphenhydramine (*Benadryl*®)
    b. heparin
    c. cimetadine (*Tagamet*®)
    d. furosemide (*Lasix*®)

26. Ideally, the best way to monitor for adequate fluid resuscitation is;
    a. heart rate
    b. blood pressure
    c. capillary refill
    d. urinary output

27. While transporting your heat stroke victim you get him all settled in and find a moment to review the sending facility's paperwork. While reviewing the labs you note the following;

Na	131 (L)	PT	26 (H)	pH	7.32	RBC	3.9
K	2.9 (L)	aPTT	82 (H)	PaCO2	20	Hgb	13.3 (L)
Cl	95	INR	1.9*	PaO2	135	Hct	38.1 (L)
CO2	18	FSP	15 (H)	HCO3	17	WBC	21 (H)
BUN	18 (H)	D-dimer	pos	BE	-8	Plat	42 (L!)
Crea	2.3 (H)	fibrinogen	165 (L)	SaO2	98%		

The heat stroke diagnosis and labs suggest;

a. your patient is hemoconcentrated with a respiratory alkalosis and multiple electrolyte deficiencies
b. your patient is experiencing the typical electrolyte disturbances associated with respiratory alkalosis and dehydration
c. your patient is in acute renal and liver failure with severe dehydration
d. your patient is in a sub-acute DIC with typical electrolyte abnormalities indicative of the precipitating diagnosis

28. Active internal re-warming of the hypothermic patient may include all of the following except;

a. ECMO (extra corporeal membrane oxygenation)
b. gastric lavage
c. hemodialysis
d. Swan-Ganz placement

29. Your heat stroke victim is showing signs of myoglobinuria. Which of the following would be least beneficial in treating this condition?

a. increase fluid administration to maintain urine output of 2ml/kg/hr
b. increase diuresis with furosemide (*Lasix*®) or mannitol
c. alkalinize the patient using $NaHCO_3$
d. hyperventilate to a slight alkalosis

30. You are called to transport a 3 month old from a very socioeconomically deprived rural area. The parents report acute onset of seizures with the infant seizing "at least a dozen times" between the 911 call and your arrival. Upon arrival, you see that the family lives in significant poverty and the infant at first glance looks underweight. The home is hot, without any form of air conditioning and a single ceiling fan in the home provides the only air movement. When asking about the infant's history they report they've never seen a pediatrician but the baby has never been "really sick". The mother stopped breast feeding the infant a month ago and began using cool tap-water two days ago for some feedings when formula became limited and more could not be purchased. While obtaining the history the prehospital team begins treatment including you partner obtaining a blood for bedside testing while another member of the team obtains IV access and administers lorazepam (*Ativan*®) to stop the current seizure. The airway is immediately controlled and high flow O_2 administered but the seizures fail to respond completely to standard dosings of lorazepam. Bedside testing reveals the following;

Na	122 (L)	pH	7.30
K	2.4 (L)	PaCO2	67
Cl	95	PaO2	112
CO2	18	HCO3	21
BUN	22 (H)	BE	-5
Crea	1.1 (H)	SaO2	89%
Gluc	42		

Which of the following is false regarding this patient.

a. febrile seizures have been complicated by malnutrition and possibly hypoglycemia
b. this patient should be hyperventilated in an effort to decrease ICP secondary to suspected cerebral edema
c. this patient may receive 3% saline at a rate which halts the seizures but does not raise the sodium greater than 4-6mEq/L with normal sodium levels not being reached in less than 48 hours
d. once seizures are controlled the sodium levels typically should not be corrected faster than 0.5mEq/L/h

31. Your patient has taken a significant overdose of nortriptyline (*Pamelor®*). You would expect which of the following?
    a. significant bradycardia with AV block
    b. apnea with secondary bradycardia
    c. treatment focused towards administering flumazenil (*Romazicon®*)
    d. behavior similar to a cocaine overdose

32. Your patient ingested a large amount of raw cocaine in an effort to hide it from law enforcement. During the arrest procedure he began to exhibit signs of possible drug toxicity and your team was called for transport. Upon arrival you are most likely to find;
    a. tachypnea, tachycardia, hypertension and diaphoresis
    b. erratic behavior, hypermetabolic state and low body temperature
    c. hallucinations (auditory and visual) with a predisposition to amorous behavior
    d. pinpoint pupils, slow or erratic breathing and tachycardia

33. Your patient ingested an unknown substance before becoming unconscious. Upon arrival you find the following ECG along with numerous unidentified pills and vodka.

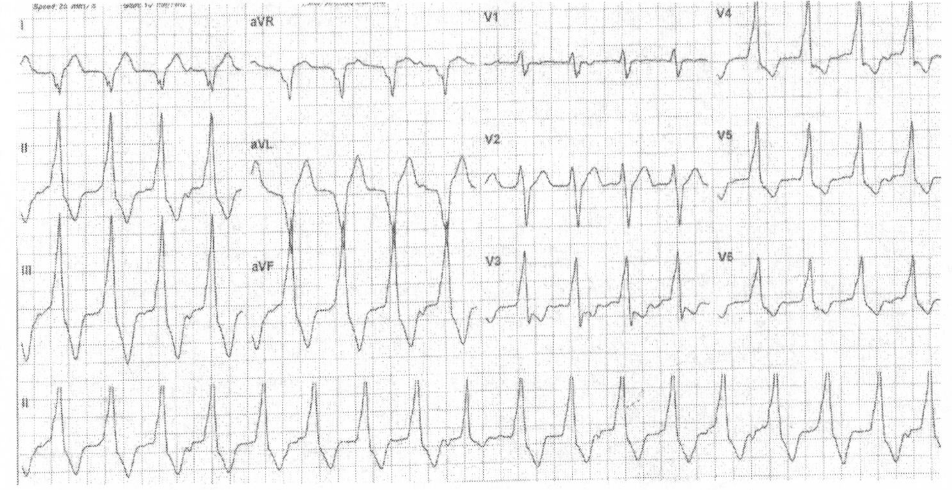

You are most likely to suspect the overdose is related to;

a. digitalis (*Digoxin®*)
b. metoprolol (*Lopressor®*)
c. amitriptyline (*Elavil®*)
d. fluoxetine (*Prozac®*)

34. Your patient was found "down" at a high-school party. At the party a small group of the boys brought "special tea". Your patient ingested a fair amount of the "tea" before collapsing. You find him disoriented and very inappropriate when rousable, warm, flushed, tachycardic with "blown" pupils. You anticipate he has ingested;

a. marijuana
b. peyote
c. MDMA (Ecstasy)
d. jimson weed

35. Your patient, a 4 year old female, ingested an unknown quantity of "grandma's candy" which she found in the bathroom. She now presents unconscious/unresponsive with a wide complex tachycardia and short episodes of torsades de pointes. Her grandmother takes the following medications; digoxin for CAD, desipramine (*Norpramin®*) for a sleep disorder, furosemide (*Lasix®*) for CHF with potassium (*K-Dur®*) and a multi vitamin with extra

iron. Which of the following choices is most likely to be beneficial in treating this patient?

a. aggressive fluid resuscitation
b. CaCl, $NaHCO_3$, Insulin, D50, albuterol and Kayexalate
c. $NaHCO_3$ with Levophed if vasopressors are needed
d. digoxin immune fab (*Digibind®*)

36. Your patient, a 76 year old male who recently lost his wife of 50 years has overdosed on his *Toprol XL®* (metoprolol). He presents with the ECG below.

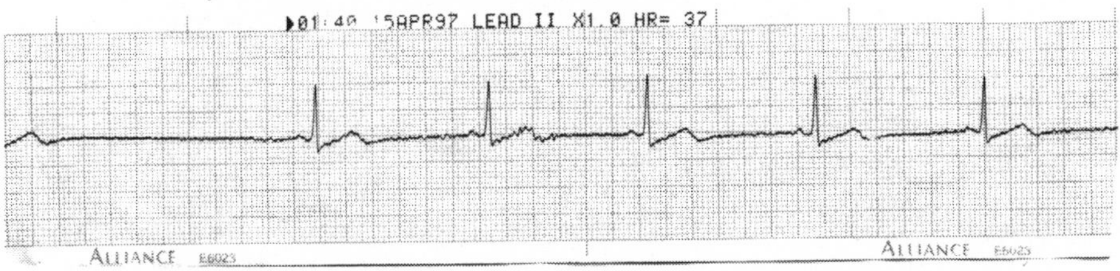

You would anticipate which of the following therapies to be unsuccessful?

a. transcutaneous pacing
b. glucagon 2-5mg IVP
c. dopamine
d. atropine

37. Your patient has ingested an unknown quantity of *Calan®* (verapamil). They present with the ECG below. Their blood pressure is very low and they are complaining of chest pain and dyspnea.

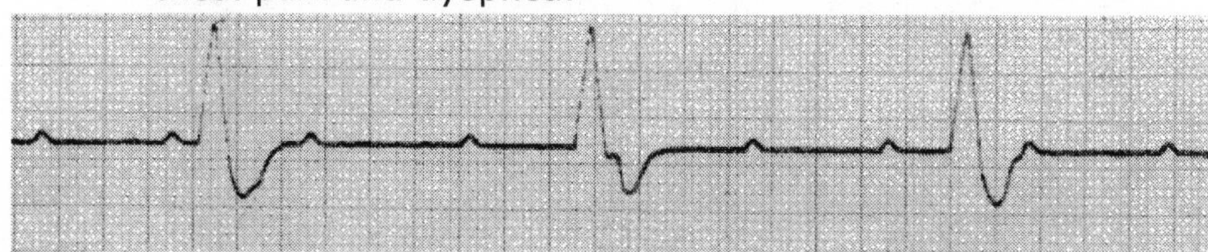

The best therapy will be administering/delivering;

a. calcium chloride
b. calcium gluconate
c. administering diltiazem (*Cardizem®*)
d. synchronized cardioversion

38. Your patient ingested a large amount of oleander leaves as a suicide attempt because he read they were poisonous. Which electrolyte should be evaluated first in relation to this toxin?
    a. sodium
    b. potassium
    c. calcium
    d. phosphorous

39. Your patient reportedly has a "toxic dig (digitalis) level". She has begun exhibiting the ECG below. Which of the following medications would not be appropriate therapy?

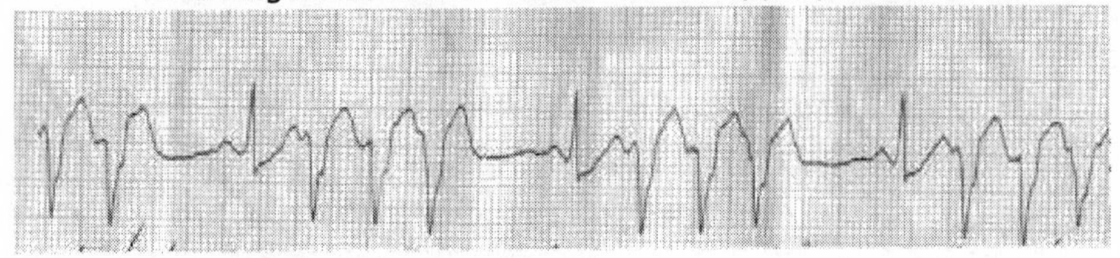

    a. lidocaine (*Xylocaine®*)
    b. magnesium sulfate
    c. procainamide (*Pronestyl®*, *Procan®*)
    d. phenytoin (*Dilantin®*)

40. Regarding the patient in the previous question, a common symptom would be;
    a. frequent urination (polyuria)
    b. orthopnea
    c. vision changes such as green and/or yellow halos
    d. report of persistent metallic taste

41. Synchronized cardioversion or defibrillation of the digitalis toxic patient will most often result in;
    a. normal sinus rhythm
    b. junctional escape rhythms
    c. continued VT/VF
    d. asystole refractory to other therapy

42. Your patient, a 56 year old male that is homeless, has ingested a large amount of an unknown substance over the last two days (per bystanders living in the same alley). He then became unresponsive. Further lab studies indicate a significant anion gap of 28 and an osmolar gap of 21mOsm/L. ABG's indicate pH 7.28, $PaCO_2$ 36, PaO_2 188 (FiO_2 0.6), HCO_3 18, BE -6. He has most likely received which toxin exposure?
    a. gasoline
    b. paint / paint thinner
    c. carburetor/brake cleaner
    d. antifreeze

43. Medications commonly used in the appropriate treatment of the patient in the above scenario would include;
    a. thiamine and 4-methylpyrazole (fomepizole) (*Antizol®*)
    b. vitamin B6 (*Pyridoxine®*) and phosphorous
    c. bicarbonate and flumazenil (*Romazicon®*)
    d. thiamine and wood alcohol

44. Your patient has received an unintentional overdose of potassium due to a medication error. Pertinent labs include K^+ 7.9 with a pH of 7.36. Their ECG is below. Which choice below is the best course of treatment for this condition?

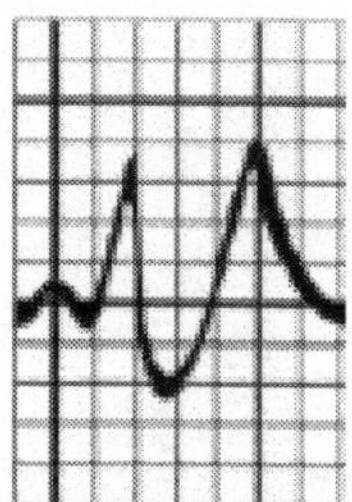

    a. hypoventilation, furosemide (*Lasix®*), sodium polystyrene sulfonate (*Kayexalate®*)
    b. CaCl, $NaHCO_3$, Insulin, Dextrose, RSI using etomidate/SUX and hyperventilate
    c. CaCl, $NaHCO_3$, Lasix, Albuterol
    d Lasix and sodium polystyrene sulfonate (*Kayexalate®*)

45. Your patient has taken an overdose of an over the counter analgesic. She is complaining of headache, "ringing in her ears", one episode of emesis and repeatedly asks, "Can I have a drink of water?" She has the following vital signs; HR 112, BP 138/88, RR 32, T 98.1°. While interviewing her she begins to experience a generalized tonic-clonic seizure which lasts approximately four minutes. Which medication did she most likely take?
    a. acetaminophen (APAP), (*Tylenol*®)
    b. diphenhydramine (*Benadryl*®)
    c, phenylephrine (*Neo-Synephrine*®)
    d. acetylsalicylic acid (ASA), Aspirin

46. Treatment of the patient in the previous question would include all of the following except;
    a. hemodialysis
    b. $NaHCO_3$ administration
    c. aggressive ventilatory support
    d. charcoal administration

47. Your 16 year old female patient decided to commit suicide by taking an entire bottle of extra-strength *Tylenol*®. She is now 30 hours post ingestion and complains of right upper quadrant pain. Her labs indicate an elevated AST, ALT and coags demonstrate an elevated PT. She is;
    a. in Stage I
    b. in Stage II
    c. in Stage III
    d. in Stage IV

48. The acetaminophen poisoning victim will most likely expire;
    a. 24-48 hours post ingestion
    b. 48-72 hours post ingestion
    c. 72-96 hours post ingestion
    d. 96-120 hours post ingestion

49. Jaundice will present in the acetaminophen poisoning victim along with other symptoms of that stage. Those include all of the following except;
    a. hepatic encephalopathy
    b. DIC
    c. nausea/vomiting & malaise
    d. peak liver function

50. Treatment of the acetaminophen poisoning victim includes all but which of the following;
    a. evaluate serum levels at least four hours post ingestion
    b. administer activated charcoal
    c. administer N-acetylcysteine (*Mucomyst*®)
    d. minimize trauma and bleeding

51. Clinical signs of ethylene glycol poisoning may include all but which of the following?
    a. elevated anion gap
    b. nystagmus
    c. fluorescent urine under UV light
    d. hypercalcemia

52. Your patient reportedly may have overdosed on cocaine. He presents with tachycardia, hypertension, tachypnea, elevated temperature and bizarre, sometimes aggressive behavior. Which of the following is contraindicated?
    a. esmolol (*Brevibloc*®)
    b. lorazepam (*Ativan*®)
    c. phentolamine (*Regitine*®)
    d. propanolol (*Inderal*®)

53. Your patient was given *Valium*® (diazepam) in a dosage of 1mg/kg by the "clinic" in a rural area for an anxiety attack. They called 911 when their patient stopped breathing. The ideal reversal agent for this overdose is;
    a. fluconazole (*Diflucan*®)
    b. fluorometholone (*FML Forte*®)
    c. flumazenil (*Romazicon*®)
    d, fludrocortisone (*Florinef*®)

54. Your patient was found apneic in an alley with pin-point pupils and 'track marks' on both arms. He is bradycardic and pulses are weak at the radial. Which of the following would help the most to prevent pulmonary edema associated with the naloxone (*Narcan*®) you are about to administer?
    a. insert an nasotracheal airway
    b. administer furosemide (*Lasix*®) prophylactically
    c. administer nitroglycerine sublingual
    d. provide oxygen via non-rebreather at 10-15lpm

55. Your patient was exposed to cyanide. The optimal antidote management is;
    a. oxygen and hyperbarics
    b. amyl nitrate and sodium thiosulfate
    c. atropine and 2-PAM
    d. physostigmine

56. Your patient was exposed to carbon monoxide. The optimal antidote management is;
    a. oxygen and hyperbarics
    b. amyl nitrate and sodium thiosulfate
    c. atropine and 2-PAM
    d. protamine sulfate

57. Your patient was exposed to jimson weed. The optimal antidote management is;
    a. glucagon
    b. amyl nitrate and sodium thiosulfate
    c. atropine and 2-PAM
    d. physostigmine

58. Your patient has received too much heparin. You could reverse the effects of the heparin with _________ but you must be especially alert to signs of _________;
    a. vitamin K ; anaphylaxis
    b. protamine sulfate ; anaphylaxis
    c. CaCl ; rebound hypertension
    d. physostigmine; acute bradycardia and hypotension

59. Your patient has a toxic coumadin level. The optimal antidote management is;
    a. oxygen and hyperbarics
    b. vitamin K and FFP
    c. methylene blue
    d. physostigmine

60. Your patient's ABG's indicate acute methemoglobinemia. You should treat this with;
    a. activated charcoal
    b. FFP
    c. vitamin K
    d. methylene blue

KEY & RATIONALE

ENVIRONMENTAL EMERGENCIES, POISONINGS & TOXICOLOGY

1. D This is somewhat a trick question. D is accurate in early hypothermia but metabolic rate does decrease in latent stages of hypothermia.

2. D Availability of glycogen is the primary limiting factor for shivering.

3. B This is why shivering can actually precipitate an AMI in some patients and the very reason most cardiac ICU's have standing orders for 'fresh hearts' to receive medication to prevent shivering post operatively. You may be thinking, "Why not just keep them warm, why drug them up?" Answer: the shivering isn't typically due to being hypothermic (although it can be), typically the shivering is a side effect of anesthesia.

4. C Simple recall information you may need for the exam.

TEST TIP
The ability to shiver is lost at 32°C

5. B With severe hypothermia, insulin begins to lose it's effectiveness and hyperglycemia is seen. This is not typically treated with more insulin as introducing it to a cold body will simply inactivate it as well. Treatment as you might imagine is re-warming of the patient.

6. D As the patient slips into severe hypothermia you should anticipate a drop in respiratory rate with CO_2 retention. This will promote respiratory acidosis. SVR will rise as peripheral vasoconstriction ensues ultimately limiting oxygenation to more tissue. This will result in a secondary metabolic acidosis. Cardiac output will eventually drop (taking cardiac index down with it) and this will further the metabolic acidosis. As the body cools, many enzymes lose their intrinsic abilities and thus stop working. Coagulation factors are disabled and clotting begins to fail, hence bleeding times will increase. This enzymatic failure is also responsible for much of the reason medications are withheld in this patient population until re-warming is performed.

7. D Delta waves are typical of WPW. The ECG below demonstrates Osborne or J-waves typical of those seen in hypothermia. Observe the deflection immediately after the QRS (beginning at the J-point). These will subside with re-warming of patient and are seldom seen in patients with a core temp above 25°C.

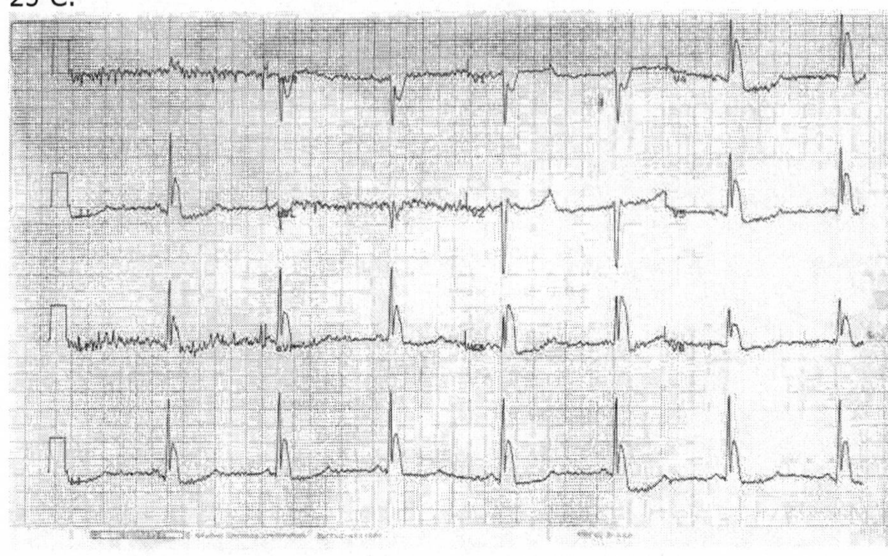

8. C	Mild hypothermia	33°C - 35°C
	Moderate hypothermia	29°C - 32°C
	Severe hypothermia	≤ 28°C

At 32°C the body stops attempting to conserve or produce heat. As such we see shivering stop, LOC and SVR will both fall, acidosis will develop and hyperglycemia occurs due to insulin's failure to function at this temperature. Below 28°C we see hypotension, ECG changes to include prolonged PRI, QRS and QT as well as the VF threshold becoming it's lowest at 22°C. PT, aPTT both increase ~50% while platelets will decrease ~40% in severe hypothermia.

TEST TIP

Mild hypothermia	**33°C - 35°C**
Moderate hypothermia	**29°C - 32°C**
Severe hypothermia	**≤ 28°C**

9. B See rationale question 8.

10. C Intubation is still recommended in verified hypothermic arrest. Cautious and gentle instrumentation of the airway is recommended.

11. B Enzymatic retardation will inhibit proper medication bioavailability and pharmacodynamics until temperature is high enough to facilitate their physiologic functions.

12. D It should be obvious that not much of anything works when treating the hypothermia patient due to the temperature. Removal from the cold environment and re-warming must be the general priority.

13. C Heat pack burns are nothing new to the literature and even more problematic with the pediatric population. "Afterdrop" or re-cooling of the core (specifically the heart) as cold blood is mobilized from the extremities during re-warming is a very real concern although it does not always occur. Rate of re-warming is increasingly important with the more hypothermic patients. Too rapid of re-warming can precipitate afterdrop and arrhythmias.

14. B Room temperature fluid is still colder than the very hypothermic patient's core. As such it will deplete the patient of much needed thermal energy via direct contact of the fluid with the tissues (conduction).

15. C "High output failure" is a very real concern in a scenario such as this. While we have no reason to believe there is underlying cardiac disease, the patient in profound heat related distress can increase his cardiac output up to four fold. With sustained CO increases such as this without relief, myocardial ischemia is very possible. While A and D are valid choices, our first priority would be C. Interestingly enough, your treatment strategy would not be significantly different in this scenario as a high output failure AMI would need aggressive fluid resuscitation, oxygenation and reduction of the myocardial workload (which cooling and fluids should provide).

16. A A low sodium would be typical of someone with prolonged fluid losses via sweating. The BUN would most likely be high and the potassium will actually be lowered as the kidneys attempt to sacrifice potassium in exchange for sodium retention. Granted the potassium levels can be much more complicated with acidosis, hyperventilation, etc. etc. etc. The CK-MM would likely be normal or high in this scenario as muscle begins to break down.

17. D Shivering and seizures are especially troublesome in heat emergencies because they generate an enormous amount of heat; increase oxygen demands to the brain ($CMRO_2$) as well as heart (MRO_2). Glucose availability actually should be dropping in this patient as his hypermetabolic state is using his stores at an accelerated rate.

18. A This sacrificing of potassium to save sodium is a normal function of the kidneys. Acute tubular necrosis (ATN) will begin to occur but it is due to pre-renal failure or inadequate perfusion. Rhabdomylosis or muscle breakdown will occur and this causes myoglobinuria, not the other way around. Finally, the PT and aPTT will elevate but this is due to the hypoperfused liver failing to produce coagulation factors.

19. D Ingesting large amounts of free water after significant diaphoresis related dehydration creates a relative hyponatremia. Heat cramps will be seen predominantly in highly exercised muscle groups.

20. A Heat exhaustion should be remembered as an elevated core temp without neuro deficits and the patient still has the ability to sweat. Once the level of consciousness changes or the ability to sweat is lost the patient has entered heat stroke.

TEST TIP
Heat exhaustion ⇒ elevated core temp, normal LOC, can still sweat

21. B A true heat stroke victim will most likely require intubation with aggressive ventilation to provide the oxygen and CO_2 exchange needed in that metabolic environment. A and C should be obvious. Phenothiazines and NMBA's will be used to prevent shivering as you aggressively cool in an effort to prevent any additional heat generation or increases in oxygen demand. Phenothiazines encourage the brain's temperature regulatory regions to become poikilothermic or assume the temperature of the outside environment (like a snake).

22. B Acute hypothermia would be a scenario such as someone falling through the ice into a lake with prolonged submersion. This scenario describes a chronic hypothermia exposure. Dry and wet hypothermia are terms I made up as distracters.

23. C Just the opposite is true, oxygen demand is exceeding supply. Respiratory alkalosis is resulting in compensation of the metabolic acidosis as well as an effort to exchange thermal energy with the external environment.

24. B These patients will need aggressive ventilation to assist with compensation, reducing workload and improving heat exchange. Using an assist control mode assures that the patient will be given a full breath with every inspiratory effort, thus reducing workload and assisting compensation. Pressure support will help overcome resistance of the vent circuit and ETT.

25. C H2 blockers are beneficial to prevent GI ulcerations secondary to the enormous stress induced by heat stroke. Remember part of the stress response will include cortisol (steroid) release in an effort to increase glucose availability. Steroids decrease PGE_2 production, which will inhibit the GI system's protective mucous synthesis.

26. D Remember that a foley is the "poor man's swan". Until the patient is producing normal urinary output they are dehydrated. While heart rate can give you rapid feedback reference the amount of compensation you are providing, urine output is "King".

27. D The electrolyte panel on the far left (Chem-7) is demonstrating typical electrolyte abnormalities associated with acute dehydration as well as significant renal failure (pre-renal very likely but true intra-renal failure will occur with myoglobinuria as well). The coag panel, second from left is demonstrating high PT and aPTT which would suggest prolonged clotting along with a positive D-dimer which suggests active clotting and clot breakdown has/is occurring. Furthermore, and more importantly, the low platelets on the CBC at far right along with the elevated fibrin split products (FSP) and low fibrinogen are the classic picture of DIC. The patient has/is clotting and has substantially used up their clotting factors. The ABG's, second from right demonstrate a partially compensated metabolic acidosis while the CBC at far right shows a slight anemia. So, while A is technically accurate, D is a more accurate representation of the situation and addresses the critical diagnosis of early DIC.

28. D Swan-Ganz insertion is simply placing the PA catheter, it does not change the core temp in any appreciable manner. ECMO, dialysis and gastric lavage are all very aggressive internal re-warming techniques that should be guided by careful clinical judgment.

29. D Myoglobinuria will result in renal failure if it is not dealt with aggressively and efficiently. Remember that myoglobin is effectively hemoglobin in the cells and it's function is to store oxygen for intracellular needs. When tissue is destroyed and myoglobin is allowed to enter the bloodstream, it is a relatively large "sticky" molecule and thus 'clogs or gums up' the glomerulus and nephrons of the kidneys. To keep the 'filters' clean we need to aggressively flood the kidneys with fluid ("the solution to pollution is dilution") and alkalinize the environment to minimize the adhesiveness of the myoglobin, allowing it to break into smaller particles for filtration and subsequent elimination.

30. A These seizures and their failure to respond to benzodiazepines is a result of the tap-water feedings and hot environment. The infant is experiencing significant hyponatremia, a hypovolemic hyponatremia at that. With the acute change in sodium, the most prominent cation in the blood, the osmolarity has shifted significantly with fluid transferring to the extravascular spaces. In this case the cerebral swelling has become significant. Correcting sodium levels is a gradual, very controlled treatment and cannot be progressed too quickly. Central pontine myelinosis (CPM) has been attributed to rapid sodium replacement but more recent studies suggest the hypovolemia accompanying the hyponatremia may be a more significant factor in CPM development.

31. D Nortriptyline (*Pamelor*®) is a tricyclic antidepressant (TCA) like amitryptyline (*Elavil*®), imipramine (*Tofranil*®) and desipramine (*Norpramin*®). Tricyclic antidepressants inhibit reabsorption of norepinephrine, dopamine and epinephrine from the reuptake channels at the synaptic cleft in a manner identical to cocaine. As a result the erratic behavior seen with cocaine overdose is frequently seen with TCA OD's as well.

32. A Remember these patients basically have a systemic over abundance of epinephrine and norepinephrine. As such they are in a hypermetabolic state which will include tachycardia, tachypnea, hypertension and hyperpyrexia. C describes more of a hallucinogenic OD, D is suggestive of an early opioid OD.

33. C Typical TCA OD's will exhibit sinus tachycardia progressing to faster rates with widening of the QRS in significant toxicity.

Eventually this will progress to VT, Torsades de Pointes and VF. You should have "Torsades de Pointes" and "overdose" linked in your mind to TCA's.

34. D Jimson weed has anticholinergic properties similar to atropine. In fact an OD of this naturally occurring plant looks just like an atropine overdose. Remember the mnemonic "Mad as a hatter, red as a beet, blind as a bat, dry as a bone, hot as a hare" to recall the clinical signs of this OD class. (Behavioral changes, flushing skin, pupillary dilation, lack of oral secretions, hyperthermia). Along with these you will find signs such as tachycardia, hypertension and tachypnea.

35. C The ECG findings along with a TCA suggest toxicity. Management of the TCA OD should be focused on alkalinizing the urine to "ion trap" the TCA for elimination. This alkalinizing also binds the TCA to albumin (protein) in the blood so that it cannot leave the vascular space to function in it's sodium channel capacity. It has been suggested that the sodium in $NaHCO_3$ is also beneficial. If hypotension is encountered, norepinephrine (*Levophed®*) is the preferred vasopressor.

36. D Current AHA ACLS recommendations still suggest administering atropine early in the management of a beta-blocker OD such as this, however it is noted that resolution of the symptoms are not expected. It is recommended based on the principle of eliminating any complicating factors such as possible increased vagal tone. The focus of beta-blocker OD management should be on transcutaneous pacing (TCP), glucagon administration with dopamine for hypotension. Isoproterenol (*Isuprel®*) was once recommended but to date no studies indicate a significant efficacy and risk of administration clearly outweighs potential benefits.

37. A The ECG rhythm may be somewhat confusing. Upon close inspection you should recognize that there is a 3rd degree AV block. There is no consistency in PRI prior to any QRS's. The relatively narrow QRS simply suggests that the ventricular pacer site is infranodal or possibly somewhere in the primary conduction pathway below the HIS bundle. Our focus with calcium channel blocker OD's is to simply over-drive the calcium channels, after all, it's a "competitive" scenario so we simply compete with more players. So why calcium chloride instead of calcium gluconate? Simple, there is a three-fold increase in the primary cation content, the number of players, (13.6mEq/gm vs. 4.65mEq/gm respectively).

38. B Oleander and foxglove are sources of digitalis. Digitalis inhibits the sodium potassium pump in cardiac cells thus creating a higher intracellular content of sodium. The sodium-calcium exchanging pump is therefore provided with a higher degree of substrate and results in an increased intracellular calcium content. In a hypokalemic environment, the interference of the sodium-potassium pump will result in more arrhythmias. Therefore the therapeutic window for digoxin narrows much more in a hypokalemic environment.

39. C Procainamide can actually exacerbate digitalis induced arrhythmias and as a general rule, procainamide is a poor choice for most arrhythmias *secondary to toxins*. Focus with digitalis arrhythmias should be on lidocaine, magnesium and phenytoin.

40. C Yellow and/or green halos are 'indications' for digitalis toxicity on certifying exams. It has been suggested that van Gogh's paintings were influenced by his possible digitalis toxicity resulting in the halos seen in many of his paintings. (i.e.- search van Gogh "Starry Night" for an example)

TEST TIP
"yellow/green halos" ⇒ digitalis toxicity

41. D Dig toxic patients respond to electricity very well. In fact you will likely convert them with minimal energy requirements. Unfortunately once converted, they frequently fail to begin a spontaneous rhythm.

42. D For purposes of the exams, the terms 'anion gap' coupled with 'osmolar gap' should immediately tip you to look for ethylene glycol, methanol, antifreeze and heavy metal poisonings. Antifreeze is a favorite of many exam item writers. Knowing how to calculate an anion gap and/or osmolar gap, while nice, hasn't proven necessary on these exams as of the writing of this text.

TEST TIP
"anion gap" + "osmolar gap" ⇒ ethylene glycol, methanol, antifreeze poisoning

43. A Treatment of these patients commonly involves thiamine, dextrose, vitamin B6 and $NaHCO_3$. Each of the choices offers one valid option but the other is invalid for B, C & D.

44. C B is a very attractive distracter. The key to it's being a distracter and not the answer is the use of SUX during the RSI. Recall that one should never use SUX in a known hyperkalemic patient. This scenario requires emergent treatment of the high potassium as indicated by the sinusoidal ECG pattern seen (as well as the documented level being very high). Emergent therapy should be focused towards rapid administration of CaCl, $NaHCO_3$, Insulin with Dextrose, Albuterol, Lasix and Kayexalate along with aggressive hyperventilation. Management may be varied slightly depending on the precipitating cause of the hyperkalemia.

45. D Classic signs of aspirin toxicity include tinnitus, thirst, headache, nausea/vomiting followed by seizures, coma and respiratory arrest. In your studies, assign "tinnitus" to "aspirin overdose".

TEST TIP
"tinnitus" ⇒ aspirin toxicity

46. D Charcoal administration can be appropriate for aspirin ingestions but recall the patient in the previous question was actively seizing. Hopefully you identified this change in level of consciousness as a contraindication to PO medications.

47. B Stage I begins 30 minutes post ingestion and lasts ~24hrs. It is characterized by nausea, vomiting, general malaise, pallor and diaphoresis. Stage II occurs at 24-48 hours post ingestion and is typified by right upper quadrant pain, elevated liver enzymes, serum bilirubin and prothrombin time. Stage III will occur from 72-96 hours and the patient typically demonstrates the peak of liver function abnormalities, return of anorexia, nausea/vomiting and malaise. Jaundice becomes apparent along with hepatic encephalopathy, DIC and death from fulminant hepatic necrosis. Finally, if the patient does not die in Stage III, Stage IV is entered around 4 days post ingestion lasting up to two weeks. This resolution period will be accompanied by return of liver function to normal baseline and the patient becomes increasingly asymptomatic. I remember these in my head as follows;

Stage I	(day 1)	ingestion to 24hrs	"flu like symptoms"
Stage II	(day 2)	24-48hrs	"owe my liver!"
Stage III	(day 3&4)	72-96hrs	"gonna die now"
Stage IV	(day 4+)	96hrs-2weeks	"I'm not dead yet*"

(-say it like the guy on Monty Python's 'In Search of the Holy Grail. It's much funnier, and easier to remember).*

48. C Stage III "gonna die now", see rationale previous question.

49. D In Stage III we see A, B & C along with peak liver function *abnormalities*.

50. B Recall that activated charcoal will interfere with the antidote *Mucomyst*®. All other therapy offered there is appropriate. Assessment of serum levels prior to four hours post ingestion will likely provide false negative values.

51. D Hypocalcemia will be seen with classic ethylene glycol (antifreeze) poisoning. The fluorescence in urine may be seen in a dark room with UV lighting if traditional antifreeze was ingested. Findings of fluorescent material on the clothing may also be discovered using this technique as well.

52. D The tachycardia and hypertension should be treated by appropriate blockade of receptors. Propanolol (*Inderal*®) is an indiscriminant beta-blocker and will inhibit beta receptors (β_1, β_2, β_3). Inhibition of β_2 receptors will inhibit large skeletal muscle vasodilation thereby *increasing* blood pressure. Using a β_1 selective beta-blocker (i.e.-metoprolol, esmolol, etc.) is appropriate along with α_1 blockade using medications such as phentolamine (*Regitine*®). Sedation using benzodiazepines is ideal for management of the psychogenic complications of these patients.

53. C A good review of standard antidotes is recommended prior to taking the certifying exams.

54. A Negative pressure pulmonary edema has been reported several times related to rapid reversal of opioids, sedatives and anesthesia. The mechanism involves the obstruction of the natural airway when the patient attempts a sudden deep inspiration. This upper airway obstruction allows for creation of a very negative pressure in the lower airways with pulmonary edema resulting as fluid is effectively "sucked" from the vasculature into the lungs tissues. An open airway is essential prior to opioid reversal.

55. B See rationale for question 53.

56. A See rationale for question 53. Recall that hyperbarics are the ultimate therapy for carbon monoxide poisonings that are significant.

57. D Physostigmine, the definitive therapy, carries with it several side effects. Administration is usually withheld for the most serious cases and minor cases are provided with supportive care only.

58. B Protamine sulfate is the reversal agent for heparin. Protamine was initially obtained from fish sperm but has since been synthetically derived. Patients with numerous allergies or known allergies to fish are most prone to anaphylaxis from protamine. Always administer a small 'test dose' prior to the full administration in hopes that an anaphylactic reaction can be reversed prior to the full dose medication exposure.

59. B Coumadin interferes with the vitamin K derived clotting factors. Reversal of this can be performed in a semi-rapid fashion using vitamin K. Vitamin K has a very high incidence of anaphylaxis associated with it's administration and as such IM administration is typically advised. IV administration is more hazardous and should be avoided. If immediate reversal of coumadin is necessary, FFP has the clotting factors that coumadin initially interferes with.

60. D Methemoglobinemia (MetHgb) is diagnosed by a methemoglobin level higher than 1.5%. This form of hemoglobin has an oxidized, or ferric iron molecule on the heme molecule vs. normal hemoglobin's ferrous iron molecule. The ferric iron cannot bind oxygen and thus a 'functional anemia' is created. Acquired methemoglobinemia commonly results from exposures to chemicals such as nitrates in fertilizers, chlorates, some fungicides, amyl nitrate (cyanide treatment actually converts the cyanide to methemoglobin allowing the body to naturally deal with the MetHgb) and many drugs to include antibiotics, analgesics and antineoplastic agents. Patients are typically asymptomatic until MetHgb levels exceed 10-15% beginning with cyanotic appearance to mucosa. Treatment is with methylene blue, an intravenous dye. You should recall that drops in SpO_2 may occur with administration of methylene blue as the dye will interfere with color reflections needed for pulse oximeters.

OBSTETRICAL AND GYNECOLOGICAL EMERGENCIES

1. The female patient will undergo a number of physiologic changes as she moves through pregnancy. Which of the following is not one of those changes?
   a. BP will decrease during 2nd trimester only to increase during 3rd trimester
   b. dilutional anemia occurs as RBC volume increases at a rate less than plasma volume
   c. heart drops and rotates rightward
   d. functional residual capacity decreases ~20%

2. The placenta;
   a. allows for direct transfer of blood between the mother and fetus
   b. is formed by chorionic and fundal plates
   c. does not develop until the 5th month of gestation
   d. is very high in thromboplastin content

3. In the pregnant female, you would expect;
   a. sphincter tone to be increased
   b. cardiac output to be increased
   c. hematocrit to be increased
   d. hormonal activity to be decreased

4. The incidence of 'failed airways' in pregnant females;
   a. is due to capillary engorgement and airway swelling
   b. occurs at a rate of 1:280 vs. 1:2230 in non-pregnant females
   c. may be very predictable using the Mallampati scoring system
   d. all of the above

5. Plasma volume in the pregnant female;
   a. increases ~40-45%
   b. is a factor in coagulopathies
   c. is a factor in airway complications
   d. all of the above

6. Attempts should be made to keep the pregnant female in a lateral decubitus position primarily;
   a. to assist with airway clearance while vomiting
   b. to assist breathing
   c. to relieve back pain/discomfort
   d. to prevent hypotension

7. Prior to transport of the pregnant female, the flight crew should consider;
   a. placement of a foley
   b. placement of at least one large bore IV of LR
   c. measurement of fetal heart tones, movement and contractions
   d. all of the above

8. The term "effacement" refers to;
   a. the thickness of uterus at the fundus, palpable through the abdomen
   b. the dilation of the cervix with full effacement occurring at 10cm
   c. the thickness of the cervix and is expressed as a percentage
   d. the position of the fetal head in relation to the mother's pubic bone

9. The sending L&D RN states that the patient is dilated 8 or 9cm, 100% effaced and the fetus is at a 0 station, the patient is still having contractions. You should;
   a. transport in slight trendelenburg position with careful attention to providing tocolytic medications
   b. transport in a slight trendelenburg, loading the patient head to tail in your fixed wing with careful attention to providing tocolytic medications
   c. transport in a slight reverse trendelenburg, loading the patient head to nose in your fixed wing with careful attention to providing tocolytic medications
   d. refuse the transport as imminent delivery is likely present

10. If the sending physician demanded transport of the above patient he would be in violation of;
    a. COBRA
    b. OSHA
    c. JCAHO
    d. EMTALA

11. The sending RN reports that the patient is being transferred because the fetus is in a transverse lie and her last delivery was breech. After manipulation the fetus returned to a transverse lie and experienced variable decelerations while in cephalic-vertex presentation. This report would suggest;
    a. possible 'short-cord' syndrome
    b. mother probably has some form of pregnancy induced hypertension
    c. mother may have gestational diabetes
    d. placenta abruptio is occurring

12. Baseline fetal heart tones (FHT's) should be;
    a. 80-100/min
    b. greater than 100/min
    c. 100-140/min
    d. 120-160/min

13. Variability should be;
    a. less than 10/min
    b. 10-15/min
    c. greater than 10/min
    d. less than 1/min

14. Variability is;
    a. the number of accelerations and decelerations occurring with each contraction
    b. the single best predictor of time to deliver
    c. the single best predictor of fetal well being
    d. the single best predictor of fetal cardiac development

15. Poor variability is caused by all of the following except;
    a. fetal hypoxia
    b. smoking by the mother
    c. extreme prematurity
    d. fetal agitation

16. Accelerations ("accels") and decelerations ("decels") refer to;
    a. rates of change in the pregnant females labor
    b. how the patient's blood pressure changes with labor
    c. how the fetus's heart rate changes with stimuli
    d. how the contraction rate changes with treatment

17. When looking at a typical fetal heart strip, the top waveform usually denotes _______ while the bottom waveform denotes ________.
    a. contractions ; fetal heart rate
    b. contractions ; breathing
    c. fetal heart rate ; breathing
    d. fetal heart rate ; contractions

18. The fetal heart monitoring strip seen below would suggest;

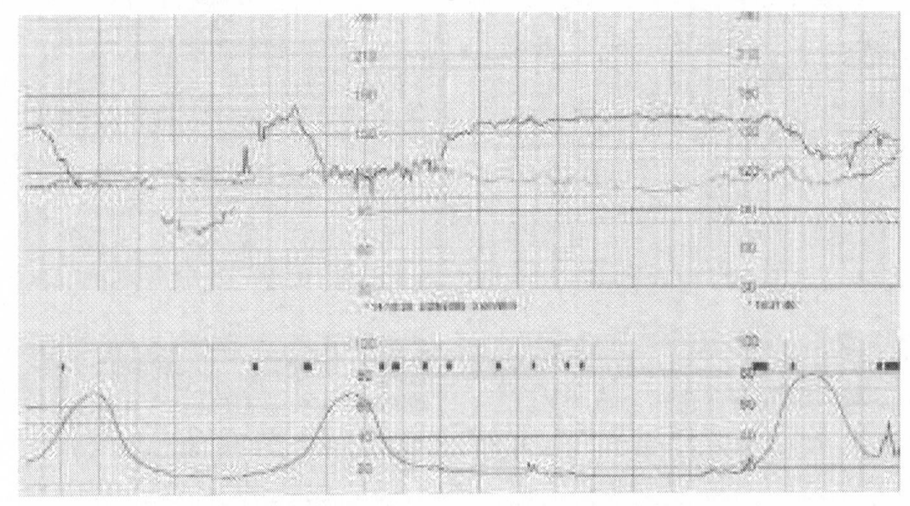

    a. early decels
    b. accels
    c. variable decels
    d. late decels

19. The fetal heart monitoring strip seen below would suggest;

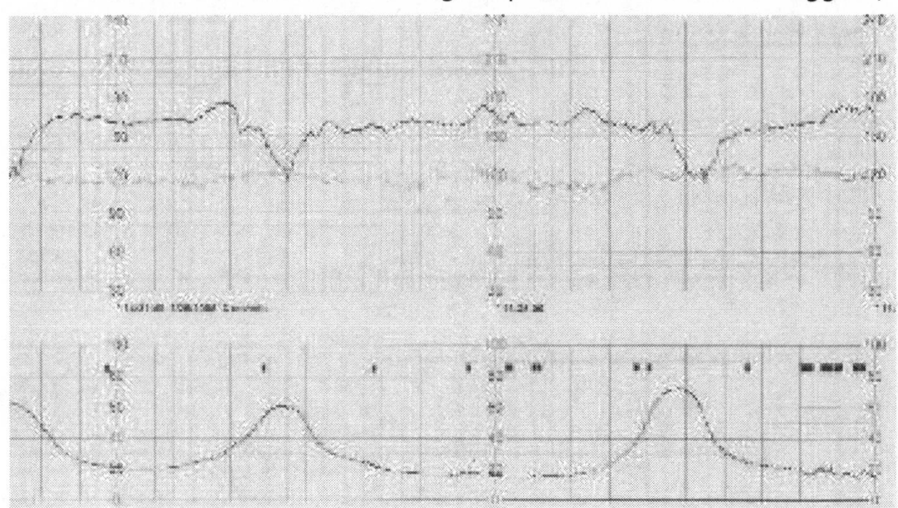

  a. early decels
  b. late decels
  c. variable decels
  d. accels

20. Good variability suggests;
  a. regular labor contractions
  b. complete cardiac development
  c. mature neurological development
  d. good placental development

21. Accelerations (accels) are typically associated with;
  a. fetal movement and response to stimuli
  b. hypoxia
  c. extreme prematurity
  d. smoking

22. Early decelerations;
  a. are very uncommon and a sign of pending newborn resuscitation
  b. commonly accepted as a vagal response to squeezing of the fetal head caused by contractions
  c. predict placenta abruptio and may be an early indicator for tocolytic administration
  d. predict a rapid/explosive delivery

23. The fetal heart strip below has a sinusoidal waveform and this suggests;

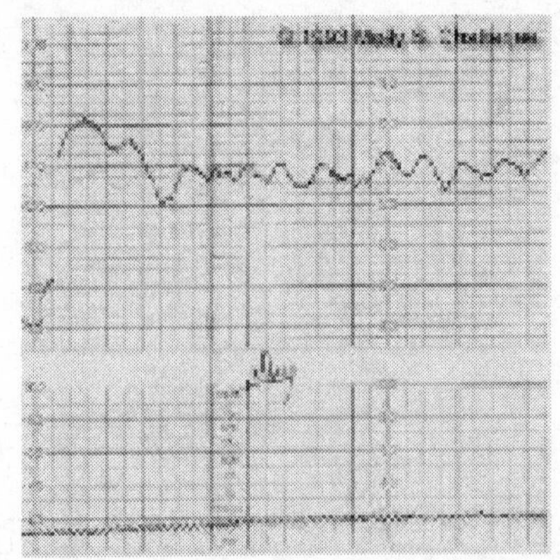

   a. that the delivery is progressing as usual
   b. that the fetus is currently being stimulated by a contraction
   c. the fetus is hypovolemic or severely anemic
   d. the delivery is distressed but fetal outcomes are good

24. Fetal bradycardia is defined as;
   a. FHT's less than 120/min
   b. FHT's less than 160/min for 5-10 minutes
   c. FHT's less than 100/min for 1-2 minutes
   d. FHT's less than 120/min for 5-10 minutes

25. Fetal tachycardia is defined as;
   a. FHT's more than 120/min
   b. FHT's more than 160/min for more than 10 minutes
   c. FHT's more than 120/min for more than 10 minutes
   d. FHT's more than 180/min for more than 5 minutes

26. The fetal heart strip suggests that the fetus is in distress and mother is experiencing tetanic contractions. Appropriate management would include;
   a. discontinue $MgSO_4$
   b. discontinue terbutaline (*Brethine*®) administrations
   c. discontinue any blood product administration
   d. discontinue oxytocin (*Pitocin*®) administration

27. The fetal distress you would anticipate to accompany tetanic contractions would include;
    a. frequent accels
    b. prolonged decels
    c. sinusoidal waveform
    d. late decels

28. The fetal heart monitor indicates the fetus is experiencing variable decelerations. Corrective action may include;
    a. administration of low-dose oxytocin (*Pitocin*®)
    b. place the patient in a supine position
    c. have the patient change positions (i.e.- roll from one side to the other)
    d. reduce fluid administration, restrict fluids

29. Signs of imminent delivery include all but which of the following;
    a. vaginal bleeding
    b. urge to push
    c. 'bloody show'
    d. crowning

30. Your patient is a 16 year old female at 32 wks gestation. She is very anxious about the delivery and pending transfer. Physically they report she is relatively healthy with 'normal vitals' and dilated to 4cm with ~30% effacement. They add the fetus is at a -3 station, membranes are intact and labor is currently controlled. FHT's have been 'within normal limits' showing fair to good variability with occasional sporadic variable decels. You would anticipate which of the following medications not to be administered prior to/or upon your arrival?
    a. $MgSO_4$
    b. betamethasone (*Celestone*®)
    c. midazolam (*Versed*®)
    d. cefazolin (*Ancef*®)

31. Your patient presented to the sending facility in obvious labor. She is an 18 year old female reporting only minimal pre-natal care (she bought a pregnancy test and talked to her mom who had 12 children). She thinks she's around 8 months but best estimates indicate she's likely around 30 weeks gestation. She was given $MgSO_4$ 2gms slow IV push followed by a drip at 4gm/hr. While transferring the patient to the airfield for loading she becomes very anxious and SpO_2 alarms indicate saturation has dropped from 98% to 56% (pleth waveform is good). Respiratory effort is good at 24 breaths per minute but rapidly increasing as she attempts to use accessory muscles. She has a functioning nasal cannula at 4lpm. Based solely on the information provided, you would most likely suspect;
    a. magnesium toxicity
    b. supine hypotensive syndrome
    c. acute hypemic hypoxia
    d. pulmonary embolism

32. Referring back to the previous question's scenario, a quick test to determine magnesium toxicity would be;
    a. cognitive exercises
    b. deep tendon reflexes
    c. pupillary response
    d. capillary refill at the periphery

33. If you suspected acute magnesium toxicity you would;
    a. administer $NaHCO_3$
    b. administer insulin and dextrose
    c. administer furosemide (*Lasix®*)
    d. administer CaCl

34. Which of the following would be an appropriate tocolytic?
    a. $MgSO_4$ 1mg IVP followed by continuous infusion at 1-2mg/hr
    b. terbutaline (*Brethine®*) 0.25mg SQ q 15min
    c. oxytocin (*Pitocin®*) 20-40units/1000ml continuous infusion at 125ml/hr
    d. hydralazine (*Apresoline®*) 2mg IVP q 5min

35. You have arrived on the scene of a car vs. immovable object crash. The driver, a male and his female passenger are both injured badly and both require blood administration. The female is obviously pregnant. Your flight program carries four units PRBC's to every scene and upon inspection you note that you have two units of type O negative and two units O positive. You should;
    a. give one unit of each type to each patient as dictated by patient condition
    b. give both O negative units to the male patient and O positive units to the female patient as needed
    c. do not give any un-crossmatched blood to the female and give the O negative blood to the male
    d. give both O negative units to the female patient and O positive units to the male patient as needed

36. Pregnancy induced hypertension (PIH) may be caused by autoregulation mismatches secondary to increased plasma volume and placental toxins. Patients most at risk for PIH include all of the following except;
    a. primagravida
    b. very young and older (over age 40) mothers
    c. women pregnant with twins
    d. hispanic women

37. PIH is a term commonly used interchangeably with pre-eclampsia and toxemia. The defining characteristics of this disorder are all but which of the following;
    a. hypertension
    b. low platelet count
    c. proteinuria
    d. edema

38. HELLP syndrome refers to a myriad of conditions which include;
    a. Hypertension, Elevated Liver enzymes and Low Platelets
    b. Hypertension, Elevated glucose and Low Protein (albumin) content
    c. Hyperglycemia, Elevated Liver enzymes and Low Prothrombin
    d. Hypotension, Elevated glucose and Low Parity

39. Your patient has been diagnosed with pre-eclampsia. You would anticipate the fetal heart tracings to indicate;
    a. variable decels
    b. accels
    c. early decels
    d. late decels

40. Initial treatment for PIH related hypertension may include all but which of the following?
    a. labetalol (*Trandate*®) 20mg IVP
    b. hydralazine (*Apresoline*®) 2mg IVP
    c. $MgSO_4$ 4-6g slow IVP
    d. terbutaline (*Brethine*®) 0.25mg SQ

41. Your patient is a 30 year old pregnant female that is 32 weeks gestation. She reports acute onset of sharp "ripping" pain across her back radiating to her chest. 12-Lead ECG shows a subendocardial stress pattern in multiple leads. She should be carefully evaluated for;
    a. pulmonary embolism
    b. placenta abruptio
    c. abdominal aortic aneurysm
    d. Braxton Hicks contractions

42. The pregnant trauma patient can lose _______ before symptoms of _________ develop.
    a. 20% FRC ; hypoxia
    b. 50% of heart rate ; hypotension
    c. 30-35% blood volume ; hypovolemia
    d. their lunch ; nausea

43. When treating the pregnant trauma patient, you should have a high index of suspicion regarding ______ and ______.
    a. speed of impact ; fetal age
    b. MOI : seat belt placement
    c. hydration status ; patient age
    d. fall distance ; assailant location

44. You should suspect placenta previa in all but which of the following patients?
    a. 23 year old female, second pregnancy, first delivered via c-section
    b. 38 year old female, third pregnancy in three years, all vaginal deliveries
    c. 16 year old female, first pregnancy
    d. 42 year old female, first pregnancy to carry to term, three prior miscarriages with D&C's to follow.

45. Placenta previa is best characterized by which of the following?
    a. painful, bright red bleeding
    b. painful, dark red bleeding
    c. painless, bright red bleeding
    d. painless, dark red bleeding

46. Placenta abruptio is best characterized by which of the following?
    a. painful, bright red bleeding
    b. painful, dark red bleeding
    c. painless, bright red bleeding
    d. painless, dark red bleeding

47. You should anticipate that the patient with placenta abruptio will;
    a. proceed into active labor that is refractory to tocolysis
    b. deliver the placenta prior to the fetus
    c. require forceps delivery as the placenta obstructs smooth transition of the fetus down the vaginal canal
    d. be at risk for fluid overload and attention to the development of pulmonary edema is paramount

48. Management of the placenta abruptio patient should include all of the following except;
    a. continuous fundal height assessment
    b. preparation for blood product administration and DIC
    c. oxytocin administration for tamponade of hemorrhaging
    d. continuous assessment of fetal movement, FHT's and contractions

49. A critical component of identifying and/or monitoring uterine rupture is;
    a. contraction timing changes
    b. serial fundal height measurements
    c. glucose measurement in amniotic fluid
    d. decreases in FHT's as labor progresses

50. Higher incidence of uterine rupture will be seen in;
    a. primagravida
    b. patients with previous spontaneous abortions
    c. patients with small body habitus
    d. patients with prior cesarean section

51. You have just delivered the head of a newborn and note a nuchal cord. Your priority should be to;
    a. suction the nose
    b. suction the mouth
    c. cut the umbilical cord
    d. administer tocolytics

52. You have identified that the patient has a prolapsed umbilical cord. The focus of your care now needs to be;
    a. inserting your hand to elevate the presenting part off the cord
    b. placing the patient in a knee-chest position
    c. placing the patient in a trendelenburg position
    d. initiating tocolytic therapy to halt labor

53. You are in the middle of a breech delivery when you realize that the head appears 'stuck' in the vaginal canal. While you can palpate the newborn's chin and lower lip with your finger via the vaginal opening, delivery appears halted. You should;
    a. perform a Trousseau's maneuver
    b. perform a Monroe's maneuver
    c. perform a Marlough's maneuver
    d. perform a Mariceau's maneuver

54. After delivery of a newborn, your patient is experiencing continuous postpartum hemorrhaging. After a vigorous fundal massage you should administer;
    a. methylergonovine (*Methergine*®) 0.2mg IM
    b. oxytocin (*Pitocin*®) 20-40 units infusion
    c. bimanual uterine compression
    d. rapid infusion of FFP and/or platelets

55. Your patient has experienced a uterine inversion during the delivery process and hemorrhage is substantial. Focus of management lies with;
    a. administration of tocolytics to promote maximum uterine contraction
    b. administration of blood products to replace lost volume
    c. manual replacement of the uterus with the gloved hand to it's natural position
    d. removal of the placenta by gentle, steady traction on the umbilical cord

56. DIC is common in the pregnant female. Focus of treatment should be on;
    a. heparin administration
    b. blood component replacement therapy
    c. minimizing trauma such as needle sticks
    d. managing the precipitating event/cause

57. During delivery your patient reports acute dyspnea which is supported by sudden drop in SpO_2 and tachycardia. All but which of the following may be a cause for this sudden change in condition?
    a. amniotic fluid embolism
    b. pulmonary embolism
    c. DIC
    d. prior pre-eclampsia

58. Your pregnant patient presented at the sending facility with complaints of headache made worse by bright lights and noise. Upon further examination she was determined to be approximately 40 pounds overweight, edematous, significantly hypertensive and urinalysis was positive for protein. During startup of the helicopter the patient begins actively seizing. What is this patient's diagnosis?
    a. pre-eclampsia
    b. eclampsia
    c. epilepsy
    d. hypoglycemia

59. The best medication to be administered to the patient in the previous question to stop the seizures is;
    a. diazepam (*Valium*®)
    b. MgSO4
    c. dextrose
    d. phenytoin (*Dilantin*®)

60. I have found this text;
    a. see rationale

KEY & RATIONALE

OBSTETRICAL AND GYNECOLOGICAL EMERGENCIES

1. C Actually the heart will be lifted anteriorly and rotated to the left along with the other changes described. In addition to those changes you should expect the heart rate to increase 15-20/min, elevation of WBC's, elevation of progesterone levels causing relaxation of sphincters and vasculature, decreased functional residual capacity (~20%) along with capillary engorgement, the two being specifically problematic during airway management.

2. D The placenta does not allow for transfer of blood between the mother and fetus (normally, some breaks do occur and are the putative mechanism behind Rh sensitization), their blood supplies are isolated, the placenta simply allows for transfer of gases (oxygen transfers at 1/5th the efficiency of the lungs) and nutrients between them. The placenta is formed by chorionic and deciduas plates and develops much earlier than the 5th month. The placenta's high thromboplastin content makes it the perfect "fuse" to trigger a coagulopathic emergency such as DIC. Recall during the trauma/hematology section we discussed thromboplastin's ability to trigger the extrinsic clotting cascade.

3. B Sphincter tone will be decreased due to the elevated progesterone levels. Hct will be lower as the dilutional anemia occurs and hormones are significantly elevated.

4. D "all of the above" questions are relatively easy. If you can identify two correct answers the answer must be "all of the above". More important is recognizing the incidence of failed airways in pregnant females. Be prepared for smaller endotracheal tubes, bleeding and increased failure rates. Have rescue airways and plans ready.

5. D The plasma volume increases a substantial 40-45% with the hemoglobin increasing only ~30%, thus the dilutional anemia. It is the plasma however, which carries the coagulation factors or the 'dynamite' if you will, for DIC. The gravid uterus promotes venous stasis in lower vasculature. Couple this increased coag factor content venous stasis and the thromboplastin 'fuse' we mentioned earlier in the placenta, and you can see why so many pregnant females go into DIC. This plasma increase is not matched by albumin production either resulting in a dilutional hypoalbuminemia. This drops the osmotic pressure gradient and results in an increased shifting of fluids extravascularly with increased edema occurring. Highly protein bound medications will also demonstrate increased potency with standard dosings as more "free drug" will be available intravascularly.

6. D The *primary* reason is to prevent supine hypotensive syndrome. The heavy uterus will impinge upon the vena cava in a supine position limiting blood return and thereby dropping preload. Most texts suggest turning to the left as the vena cava lies predominantly to the right of the aorta and spinal column. Ultimately turning to the right may be acceptable as is simply displacing the uterus to one side manually should supine positioning be required for some other reason.

7. D Due to the enlarged uterus and decreased sphincter tone, voiding is frequent and difficult to manage, foley placement is helpful in these regards as well as essential in managing fluid resuscitations. Transport crews should consider at least one if not two large-bore IV's of LR. Continuous monitoring of FHT, movement and contractions is essential in the actively laboring patient. Placement of an NG tube should also be strongly considered due to; high potential for intubation, delayed gastric emptying during labor, increased intragastric pressures and reduced lower esophageal sphincter tone.

8. C 'D' refers to "station".

9. D Hopefully you've identified that a cervical dilation of 8-9cm is almost fully dilated, coupled with being fully effaced means that cervix isn't holding anything back. The station of 0 means the fetus is 'cocked and ready to go' as the head has engaged the exit. This baby will begin delivering any moment, especially with mom still actively laboring.

10. D EMTALA or the Emergency Medical Treatment and Active Labor Act was part of the Consolidated Omnibus Budget Reconciliation Act (COBRA) legislation of 1986 and was written to prevent "patient dumping" or the act of refusing to care for financially indigent patients. Basically the law states;

A participating hospital (one that receives payment from Medicare) must provide an appropriate medical screening exam to anyone coming to the ED seeking care to determine if they have a medical emergency or active pregnancy labor.

If the hospital determines that the individual has an emergency medical condition or is actively laboring, the hospital must treat and stabilize the emergency medical condition prior to transfer.

11. A Variable decelerations are typical of a 'cord problem'. The umbilical cord is being stressed in some manner such as impingement, nuchal cord, short cord, prolapsed cord, etc.

TEST TIP
variable decels ⇒ "cord problem"
variable decels ⇒ "V or W shaped"

12. D Definition

TEST TIP
FHT's are normally 120-160/min

13. B Variability refers to how much the fetal heart rate is varying on a second-to-second basis. Variability is the single best predictor of fetal well-being and suggests that the neurological system is developed with the fetus responding to external stimulus.

TEST TIP
"variability is the single best predictor of fetal well-being"

14. C See rationale question 13.

15. D Any factor which causes hypoxia or immature development will dampen the variability. Agitation, such as that seen during delivery should increase variability of the normal fetus.

16. C Monitoring of the FHT's will demonstrate generalized increases (accelerations/accels) and decreased (decelerations/decels) in rate depending upon the situation. Generally speaking, accels are good, decels are bad. Various types of decels must be reviewed and understood.

17. D This is the standard most commonly seen but variations can exist. Commonly a third tracing is seen which will represent the mother's heart rate as well.

18. D The decelerations noted in the fetal heart rate (upper tracing) are occurring a significant time after the peak of the contraction (lower tracing) is seen below. Late decels are indicative of uteroplacental insufficiency and commonly occur with PIH, DM, smoking and late deliveries. Essentially consider any cause that might make the placenta unable to keep up with the demands of a fetus. DM and late deliveries typically involve larger fetus's, therefore they need more substrate/oxygen which a normal placenta can't keep up with. PIH causes a 'tighter set of pipes' if you will. Uterine blood flow is dependent upon systemic BP, uterine venous pressure and uterine vascular resistance (how tight are the pipes). With decreased uterine flow we see decreased oxygen delivery to the fetus. Smoking should be obvious, the increased carboxyhemoglobin levels in the fetus do not facilitate good oxygenation. Fetal hemoglobin is structurally different than normal hemoglobin and as such can carry 20-50 times more oxygen with their Oxyhemoglobin dissociation curve shifted slightly left.

TEST TIP
"late decels" ⇒ uteroplacental insufficiency
"late decels" ⇒ PIH, DM, smoking, late deliveries

19. C Variable decels are commonly 'V' or 'W' shaped. They commonly reflect umbilical cord related problems. Issues such as nuchal cord, short cord, prolapsed cord and entanglement are possibilities to consider. Once the cord is stressed, impinged or stretched, vascular flow through the cord is significantly altered and hypoxia ensues. Variable decelerations typically occur during contractions which is very intuitive when you consider the physical mechanisms involved. Obviously the contraction and changing the infant's physical relationship to the placenta or cord is causing the distress.

20. C See rationale question 13.

21. A See rationale questions 13 & 15. If you were *'shootin dark'* on this question a test taking strategy to use would be identifying which answer is not like the others. B, C and D are all bad things, A is not.

22. B Early in the engagement phase the head will be squeezed as it begins the process of dilating the vaginal canal. The well-developed parasympathetic nervous system will respond to the pressures with bradycardia. Once the head has moved into the vaginal canal this may or may not subside.

23. C Sinusoidal waveform is a very bad sign. This typically indicates extreme neurological problems commonly associated with accidental tap of the umbilical cord during an amniocentesis, fetomaternal transfusion or placenta abruptio. Delivery of the fetus is priority and prognosis is poor.

24. D It's a definition.

25. B Another definition.

26. D Tetanic contractions are continuous or 'back to back' contractions. The fetus is being squeezed continually. This is most likely when medication such as oxytocin is being administered.

27. B With tetanic contractions, stress upon the fetus will result in an increased oxygen demand while the tense intrauterine environment will inhibit blood flow via the umbilical cord, aka- the oxygen supply. Fetal hypoxia will result and be demonstrated by prolonged decelerations, late decelerations or fetal bradycardia. (See strip below)

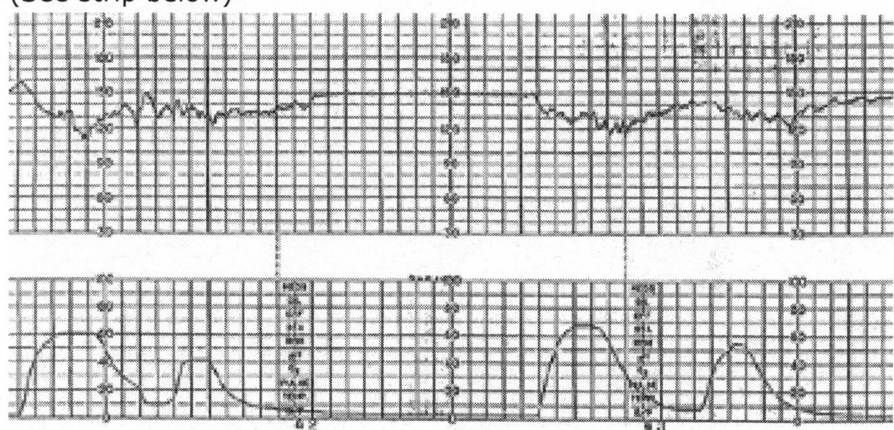

28. C Recall that variable decelerations are typically associated with 'cord problems'. Maneuvers as simple as rolling the patient to the opposite side may be all that is required to get the infant off their cord or facilitate the mechanical changes needed for better perfusion via the cord.

29. C "Bloody show" refers to the bloody mucous plug that obturates the cervical opening, being discharged as the cervix begins to dilate. This is a very early sign of labor and should occur long before imminent delivery is apparent.

30. C As a general rule, you should stay away from benzodiazepines in pregnant females, especially approaching delivery. Their ability to cross the placenta is very good and can precipitate a 'floppy' newborn with inadequate respiratory effort.

31. D Sudden onset of respiratory distress with poor SpO_2 that is unresponsive to supplemental O_2 should always ring the "PE alarm" in your brain. Add in the fact this patient is "pregnant" aka - "hypercoagulopathic" and you will stay ahead of the game hopefully. Magnesium toxicity is a concern, she did receive a good loading dose and her infusion is in the higher range. However there is no mention of DTR's, sedation or hypotension in the scenario. Look for those three in combination on OB questions pertaining to magnesium toxicity.

32. B Typically the best, fast test for Mg toxicity is DTR's. Refer back to the last question for typical presentation of such.

33. D The key here is "acute" toxicity. Acute toxicity must be managed aggressively with calcium. Chronic magnesium toxicity such as that seen with prolonged infusions and such can typically be managed with loop diuretics because magnesium is efficiently eliminated via the renal system.

34. B If I got you with A, look at the dosage again. It says 1 milligram, not 1 gram. Simple questions like these are painful at the end of a long exam because your tired brain reads what it wants to read. Use caution.

35. D Remember that O negative is the universal donor so it's safe to give that blood type to anyone. Furthermore, the primary reason behind Rh typing is to prevent sensitization of the female to Rh+ blood type. Once she becomes sensitized, she is more likely to spontaneously abort an Rh+ fetus, thus we do not want to administer O positive blood unless we absolutely must. In the event that does occur, we need to be sure that we report the transfusion so the patient can receive RhoGAM to prevent the maternal sensitization to future Rhesus positive fetus's.

36. D Actually African-Americans are most likely to develop PIH during pregnancy, which is somewhat intuitive considering the African-American predisposition to hypertension in general.

TEST TIP

PIH is most likely to develop in;

- **African-Americans**
- **first time pregnancies**
- **very young & much older women**
- **women pregnant with multiple fetus's**

37. B While low platelet count is a defining component of HELLP syndrome, a specific form of PIH, the other three components are found in most all forms of PIH.

38. A HELLP is a specific variant of PIH, which typically becomes most evident in the third trimester and occasionally post-partum. Its clinical presentation is very similar to the other PIH variants including headache, photophobia, blurred vision, malaise, nausea/vomiting, edema and hypertension.

39. D Late decelerations are tied most closely with PIH and pre-eclampsia diagnoses. The late deceleration occurs after the contraction and signifies the fetus's inability to recover from the stress of the contraction itself because of the associated uteroplacental deficiency can't keep up with the fetus's demands.

40. D Terbutaline (*Brethine*®) is used commonly in obstetrics as a tocolytic agent, not for hypertension or PIH management per se. All other agents treat hypertension via various mechanisms.

41. C 50% of females under age 40 with a AAA are pregnant. Increased aortic pressure from the distal impingement caused by a gravid uterus along with increased circulating volume and increased aortic ejection pressures all contribute to this diagnosis. Remember that the third trimester is the period of maximal hemodynamic stress. Significant AAA can progress into the ascending aorta resulting in AMI as well.

42. C Due to their significant volume increase, a large volume loss must occur before typical signs of hypovolemia. In this respect, treat pregnant patients like children. Keep them 'tanked up', anticipate losses and do not get behind.

43. B Very low speed accidents can have grave outcomes if the mechanism of injury is just right or a seat belt is poorly placed.

44. C The other three have history that suggests uterine scarring or insufficient healing time between pregnancies.

45. C This is your telltale for placenta previa.

TEST TIP
"painless bright red bleeding" ⇒ placenta previa
"painful dark red bleeding" ⇒ placenta abruptio

46. B This is the sign for placenta abruptio.

47. A Once blood contacts the uterine wall, labor is almost unstoppable. Blood is very stimulating to the uterine wall.

48. C Administration of oxytocin (*Pitocin*®) would further aggravate placenta abruptio as delivery would be accelerated. Ideally the abruptio patient should deliver via cesarean section in the relatively controlled environment of an operating room.

49. B Serial fundal height must be monitored with uterine rupture.

50. D Prior cesarean section causes a physical defect in the uterine wall that may become the 'chink in the armor' if you will.

51. C Relief of the nuchal cord is essential prior to the next contraction and takes priority. Whether you are able to move the cord over the shoulder or it requires cutting, the cord is priority.

52. A All of the answers presented should be performed but priority lies with assuring adequate circulation via the umbilical cord.

53. D Mariceau's maneuver involves supporting the infant's body with one hand and forearm while applying gentle pressure to the mother's suprapubic area with your other hand. As pressure is applied, the arm supporting the infant, lifts the infant's body flexing the neck. This maneuver results in the newborn appearing to do a 'flip' onto the mothers belly with the newborns feet towards mom's head.

54. B Oxytocin (*Pitocin*®) is your first drug of choice, typically followed by *Methergine*®. Bimanual uterine compression may be required in the most severe cases. FFP and/or platelets would likely be part of the fluid resuscitation but getting the uterus to 'clamp down' is priority in this emergency.

55. C Regardless of smooth muscle contraction, the uterus cannot adequately reach hemostasis without being in its natural position. The uterus must be manually pushed upward and re-inverted. Do not administer oxytocin (or any other treatment that will promote uterine contraction, i.e.- baby suckling, nipple stimulus, etc.) before this is done or re-inversion may prove impossible. The uterus must remain 'loose' until back in proper position, then oxytocin would be indicated if bleeding didn't halt naturally.

56. D This principle is actually the only consistent management strategy suggested by all schools of thought and for all patient populations. In this case the delivery needs to be completed and vaginal hemorrhaging stopped. Once causation is identified and managed, then the most common management strategy with pregnancy induced DIC is blood component replacement therapy.

57. D You should have immediately suspected pulmonary embolism. Amniotic fluid embolism (AFE), while fairly rare, does still occur and is relatively fatal. Keep in mind AFE can occur postpartum with most deaths occurring within five hours of delivery. AFE is identified by a process of elimination and mimics anaphylaxis very well. DIC can precipitate significant occlusion of the pulmonary tree however this typically takes a little longer. Pre-eclampsia hopefully was a lousy distracter.

58. B Her history and physical findings are typical of pre-eclampsia. Once she seized she became officially eclamptic.

59. A Magnesium is commonly used immediately to prevent further seizures but this appears to prevent cerebral vasospasm more than anything. While the patient is seizing the stress and hypoxia induced on the fetus are overwhelming. Remember a fetus has essentially no oxygen reserves at all. Immediate termination of the seizure with a benzodiazepine is most appropriate, followed by $MgSO_4$. If delivery occurs shortly thereafter, be prepared for a "floppy" baby needing respiratory support.

60. I hope you found this text beneficial. Please feel free to contact me at www.theresqshop.com to voice concerns or provide suggestions for improvements. I can't express enough my admiration for anyone willing to take the time to teach themselves. Good luck on your future exams and be safe. You have my sincerest best wishes. Thank you.

Stay safe

Stay sharp

(ACE SAT)

57300892R00177

Made in the USA
Lexington, KY
11 November 2016